WHY THE HECK AREN'T WE ALREADY DOING THIS STUFF?

12 New Economic Policy Ideas Even a Congressman Could Understand

Al Lewis

CONTENTS

FOREWORD 1

INTRODUCTION...5

Part One
Ideas Requiring One Lousy Little Sentence to Implement 27

Chapter One
Pennies from Heck: How America is Being Crucified on a Cross of Zinc 31

Chapter Two
Reducing the Cost of Being Poor..43

Chapter Three
Behind the Green Door..49

Part Two
Micro-Heckomic Ideas Requiring the Government to Actually Lift a Finger 57

Chapter Four
The Shopamerica Gift Card:
The Grandmother of All Gift Cards...59

Chapter Five
Starting a Kinder, Gentler Trade War..73

Part Three
Eliminating middlemen, those leeches who suck the lifeblood out of an economy at taxpayer expense to further their own evil plans, such as me...81

Chapter Six
Time to Blow Up the Real Estate Brokerage Cartel
(And Find A Way To Notice Those Hardwood Floors All By Ourselves) 85

Chapter Seven
Your Medicare Tax Dollars at Work. Do You Mind If I Step Inside? 109

Part Four
Ideas Guaranteed to Attract Opposition from People Who Don't Even Know Any Middlemen 123

Chapter Eight
A Kinder, Gentler Death Panel...127

Chapter Nine
A Kinder, Gentler Obscenely High Gasoline Tax.............................147

Chapter Ten
Reducing the Cost of Being Poor Even More....................................169

Chapter Eleven
Send Us the Richest Refuse of Your Teeming Shores.......................181

Part Five
One Final Step to Kick the Economic Methadone Habit.............193

Chapter Twelve
The Body Count:
Why Measurement Matters...195

Part Six
Inquiring Minds Want to Know
How to Win the $1 Million Prize..207

BIBLIOGRAPHY...215

ACKNOWLEDGMENTS...217

About the Author...221

FORWARD

A TYPICAL IDEA YOU COULD HAVE THOUGHT OF. BUT YOU DIDN'T AND THAT'S WHY YOU HAD TO BUY THIS BOOK INSTEAD

The nice thing about a self-published book is that you don't have to bother with things like rejection letters, agents with 200 other authors to service, publication schedules with 12-month lead times, unreasonable deadlines, anal-compulsive editors, six-figure advances, *Today Show* interviews and movie rights.

But the biggest advantage is that you can add whatever copy you want between printings. No one at a "vanity press" cares what you say in print, and it only costs you about twenty bucks to change the interior. You can even make mistakes that the aforementioned anal-compulsive editor at a so-called "real" publisher would pick up but that most readers would never have noticed. For instance, the lead-in to a book is called a "foreword," not a "forward."

That flexibility means that the following idea did not appear in the previous "edition" (meaning the first hundred copies that got printed), mostly because I hadn't thought of it yet. I will be adding ideas to every "edition" as soon as I think of them or someone else suggests them, because most American citizens would agree that we could use more ideas, sort of like most Krypton citizens would have agreed that they could have used more rocket ships.

Imagine if there were a way to revitalize downtowns, reduce crime, increase retail sales, reduce seasonal depression, boost tourism, save energy, and get more people to exercise. And, except for the $20 I just paid, it wouldn't cost anyone a dime. Well, there is a way to do all that.

All these benefits would be ours if we just move the boundaries of each time zone about 300 miles west. The sun would set an hour later all year-round in the four 300-mile-wide swaths of the country currently in the eastern sections of their respective time zones, like the Northeast, where the sun sets ridiculously early right now. Those regions would then become part of the western regions of the new time zone configurations, and enjoy much later sunsets. This proposed time zone adjustment would encourage many of us to get outside and do stuff, some of which would involve spending money and hence stimulating the economy, on warm summer evenings.

Don't take my word for it. This is exactly what happens in Europe, where the cities hum with foot traffic all evening and summer daylight lingers until close to 10 PM.

Let me put it another way. Here in Boston, even with Daylight Savings Time, on June 21 the sun rises at 5:07 AM and sets at 8:24 PM. It's not just all that sunshine getting wasted in the morning. That doesn't affect the economy. What does affect the economy, not to mention our enjoyment of life, is what happens quite predictably at the other end of the day – the reality that the sun sets here earlier than any other major city in the world at this latitude.

A brief history lesson also helps to show what a dumb idea it is to keep the current time zones in place.

Throughout most of history, every town was basically its own time zone, with local time established according to when the sun was directly overhead. (Hence the expression "high noon".) Since travel between towns hardly ever took place in a time-sensitive way, there was no need to standardize times between towns.

Besides, our ancestors rose and retired with the sun anyway. That's all they needed to know about time. Needing to know

2

the time in their own town rarely came up (and when it did, well, that's why churches have bells and factories had whistles), while needing to know the time in other towns was never a pressing issue for anyone.

On those rare occasions when it mattered that a time in one place didn't match the time in another, people improvised. Like when a sheriff would tell no-good varmints that he wanted them out of town by sundown, it wasn't because he was trying to sound like Walter Brennan or Foghorn Leghorn. It was because if the varmints in question were from a different town, which varmints usually were in those days, they were also from a different local time zone. Providing even the most punctual no-good varmints with a specific time-of-day deadline to get out of town would just have created confusion, not to mention resentment of the aforementioned sheriff, possibly leading to a fatal gunfight involving Eric Clapton.

But literally and figuratively, times change, and as communications and transportation became faster, it did begin to matter that local times weren't matching. Regulated time zones were first proposed in the 1880s by those beacons of corporate civic-mindedness, the railroads, run by regular guys like Jay "The Public Be Damned" Gould, in order to standardize their timetables. (It is one of the ironies of American history that, even though it was their invention that spawned the idea in the first place, nowadays trains are about the least likely things to arrive on time. Thanks to the soaring rate of C-sections, even babies arrive more punctually.)

The time zones were set in an era when half the population farmed and the light bulb was still a curiosity. Consequently, you won't be surprised to learn that in those days, average bedtimes were two whole hours earlier than they are today, meaning that the current time zones were set when we all had much different daily rhythms.

And yet we are still using the same ones used in the 1880s, even though once again, literally and figuratively, times have changed. Yes, we added Daylight Savings Time a century ago. And there are a number of people – enough to generate 286,000 Google hits – who want to move to year-round daylight time. The irony is, that is kind of a dumb idea. In the winter, there is only so much daylight to go around. A move to year-round daylight time would mean that instead of going to work in the light and coming home in the dark, we'd be going to work (and also, in many places, waiting for school buses) in the dark and coming home in the light. This idea wouldn't do the first thing to address the wasted summertime sunlight.

In fact, given a choice between adding 3 more months of daylight time in the winter, and moving another hour ahead for the three summer months (Memorial Day weekend through Labor Day), I'd opt for the latter because summer is where the benefits accrue. But the simple one-time time zone adjustment accomplishes that without the extra seasonal adjustment.

Every other government policy from long ago, even the subsidy for mohair farmers, has been updated. Why should time zones be any different? Put another way, who should set your bedtimes? You, or a group of 19th-century robber barons?

Are you convinced yet? Good. Now go onto the www.whytheheck.com site, find the time zones posting, and send it to your Congressperson. That's the way *Why the Heck* works. No one is going to listen to me but a lot of people will listen to us.

INTRODUCTION

You are about to be engrossed by a compelling tale of love, betrayal and economics, but mostly economics. Don't worry if you are not an economist. The economics part is quite accessible to non-economists. This is partly because *Why the Heck Aren't We Already Doing This Stuff?* avoids things that make other economics books real turn-offs, such as equations, numbers and actual facts. But mostly it's because I am such a great writer. Don't take my word for the latter – just look at the evidence:

- I am blessed with such effective persuasive powers that I was once able to convince a resort to sell me a timeshare;

- My first book got eight five-star reviews on Amazon, including four from people I don't even recall having slept with.

The bottom line: The economic analysis is presented clearly enough that readers who understand it can probably continue to live independently for at least a few more years. And the ideas are simple enough that *Why The Heck* can present an entire program to increase our standard of living – in responsible, equitable, sustainable, ways – in less than 220 pages. Brevity is paramount because leading environmentalists are now urging authors to shorten their books, in order to conserve valuable resources, such as paper, ink, and semicolons. So you see, just by buying *Why the Heck* – even before you read it – you are helping to preserve the environment by reducing the average size of books sold in America. That means the more copies you buy, the more trees you save. Do the math if you don't believe me.

Why The Heck is as clear and simple as economics gets, because it is not about obscure topics like the "yield curve." Not only do we avoid talking about the "yield curve," but just to be on the safe side, we don't even talk about "yields" or "curves" separately. Rather, *Why the Heck* is about new "outside the box" ideas – ideas that no one has thought of until now – to sustain economic growth and make America #1 again in standard of living. Ideas that will make you say: "Why the heck aren't we already doing this stuff?"

Speaking of living standards, here's the problem: You were probably brought up being told that we were the richest country on earth. Maybe that was true back in the days of *Gilligan's Island,* where roughly one in every seven inhabitants was a millionaire (or his wife). But now fast-forward from *Gilligan's Island* to *Survivor,* where the islands are so poor that the set designers can't even afford *light bulbs.*

You see my point? That's okay. Neither do I. But one way or the other, our standard of living has fallen to roughly 20th in the world now. We are even behind Switzerland. *Switzerland.* What a bunch of wimps. Whereas the United States makes manly products like beef and guns and forklifts and Jumbotrons and Quentin Tarantino movies, Switzerland's major industries are tourism, watches, trains, cocoa powder, sheep, milk chocolate, dark chocolate, Nestle's Quik, Ovaltine, and yodeling.

Why The Heck isn't just a new kind of "outside the box" economics. It's a new kind of book, for a new kind of reader. Specifically, *Why The Heck* is not just about the ideas already in this book — it's about your own outside the box ideas to improve our collective well-being. We want readers not just to read *Why The Heck,* but also to contribute their own ideas for subsequent editions. No, we're not asking for charity here – we'll pay you for 'em. So shut up and keep reading.

The reason I am asking for more ideas is that the type of economy we have now, and will have for the foreseeable future, does not sustainably increase living standards for the average person. Most Americans felt better off in the last century. True economic growth that makes everyone better off (without degrading the environment for future generations) won't happen without fundamental, creative ideas to improve the way the economy functions and foster the next generation of innovation, production and — then and only then — sustainably higher levels of consumption. Until such ideas are implemented, improvements in our standard of living are make-believe, and reflect only a transition from unsustainable consumer spending generated by the "economic heroin" of illusory private wealth to unsustainable Government spending to provide economic stimulus.

This government stimulus has created a "Methadone Economy," spending so much taxpayer and borrowed money that the public sector is now a larger contributor to total economic demand than the business sector, comrade.

Why do we call it a Methadone Economy? Because Methadone doesn't provide the highs of heroin, but it can nonetheless be addictive. The best way for the Government to prevent a Methadone-like addiction to ongoing stimulus is to implement sound, creative Outside the Box policies. The first step in implementation is identification of those policy ideas, and that's where www.whytheheck.com comes in. The best Outside the Box idea submitted to www.whytheheck.com could win $1 million. There are smaller prizes too. Any idea good enough to be published earns $100. And even one-paragraph comments on other ideas, if published, earn $20. All your questions about the prizes are answered in Part Six. Here, in order, are the other most common questions and answers about *Why The Heck*:

- How do I know whether I will like this book?
- How would I know if I can "think outside the box"?
- How do the *Why The Heck* ideas differ from macroeconomics?
- How did we get into this Methadone Economy and how do we get out of it?
- Why did you selflessly start this website and fund this prize? In other words, why are you such a great guy?
- What constitutes a good idea?
- Are you really offering a million dollars for the best idea?
- What are the best ideas so far?
- What are the worst ideas so far?
- What has been the reaction of the actual people in Washington?
- Are you looking for Democratic-type ideas or Republican-type ideas, neither or both?

Let me take a shot at answering those for you.

Q: How do I know whether I will like this book?

You will like *Why The Heck* if you have a sense of humor, are open to new ideas, and can think outside the box. In other words, my accountant *hates* it. (If your accountant likes it, you should probably switch accountants.)

Q: How would I know if I can Think Outside The Box?

Most people can't Think Outside The Box. How do we know this? A recent study showed that this simple four-word phrase has now overtaken "let's touch base" as the #1 most hated cliché in the workplace. Obviously, if people could do it, they wouldn't hate it. (For instance, "Take a deep breath" doesn't even make the Top 100 hated-cliché list, because nobody hates breathing.) Maybe you're the exception. The only way to find out is to take this quiz.

In each pairing, the first character is "inside the box," while the second character is his/her/its/their "outside the box" counterpart. Give yourself one point each time you identify more with the second character.

"Inside the Box"	"Outside the Box"
Captain Ahab	Captain Crunch
Smokey the Bear	Tony the Tiger
Bronte Sisters	Blues Brothers
The Monster that Ate Pittsburgh	Herman Munster
The Man from U.N.C.L.E.	Uncle Fester
Tarzan	Magilla Gorilla
Charley Manson	Charley the Tuna
Investigators, Watergate Scandal	Investigators, Monica Lewinsky Scandal
People Who Have Been Abducted by Aliens	E.T
Rin Tin Tin	Astro

HOW TO SCORE THIS QUIZ: Congratulations! You passed! Yes, I know we didn't ask you what your score is. Frankly, we don't give a damn what your score is – you still passed. Why? Because you've read this far, that's why! That means you're willing to believe that there is any value whatsoever in scoring some stupid quiz containing random lists of characters, most of which are fictional, and almost a quarter of which are talking animals from the 1960s. You are not only far enough outside the box enough to appreciate a good, creative pop economics book, but you are also far enough outside the box that, depending on your state of residence, your legal guardian may be required to alert the authorities.

Q: How does *Why The Heck* economics differ from macroeconomics?

While *Why The Heck* is about the economy as a whole, it overlaps only slightly with macroeconomics, which deals largely with government spending. Not just government programs, but also bailouts and stress tests and companies "too big to fail" and a whole lot of other stuff which, bottom line, involves giving away or "investing" taxpayer money. *Why The Heck* is none of that. *Why The Heck* is about applying imaginative solutions to our problems rather than throwing money at them.

Why the Heck's brain trust (me) thinks government spending totally lacks imagination. Any idiot can make an economy grow in the short term by borrowing money from China and promising that our children will pay it back. Obviously the Government can't borrow its way to prosperity in the long term any more than you and I can. The deficits will need to be wound down. Government spending must ultimately be replaced by more sustainable economic policies...and that's what *Why The Heck* is all about, specific policies to:

- Increase confidence in the Government, which will increase spending;
- Sustain and grow the economy without government spending or tax cuts;
- Create market conditions friendly to new investment for long-term growth; and

- Smooth the functioning of the economy by addressing "drags" so that maintaining growth becomes easier.

Why The Heck also espouses the philosophy that the Government should expand from its traditional roles – ensuring order, investing in infrastructure, providing a social safety net and starting unnecessary wars with countries ruled by tinpot dictators who had nothing to do with 9/11 but who once dissed a certain president's father – into another role that has traditionally belonged to the private sector: creating efficient markets out of inefficient ones. Several chapters propose free market improvements that in theory should happen without government intervention, but that for a variety of reasons haven't. Creation of efficient markets helps to raise living standards, and ultimately that's what will make us #1 again, or at least #19.

The reason you've never heard of this approach before and the reason the field didn't exist before was that there was no need for it. The economy grew on its own. And it was not hard to get people to spend money, customarily before they even earned it. That's why God invented the credit card.[1]

Starting in 2008, though, people began managing their money like responsible adults, which is about the worst thing that could happen to a consumer-driven economy. For instance, if we learned anything from the fiasco that was the 2008 Stimulus Check mailing, it's that we really do need to learn how to get people to spend money, and spend it productively, instead of

---

1. OK, unlike the large majority of the other sentences in this book, that sentence is not technically true. Historians have documented that the credit card was conceived in 1950 by Frank McNamara, after he took his wife out for dinner but failed to notice he had left his money at home. Historians have also documented that Mr. McNamara slept on the couch that night.

saving it and/or spending it on gasoline, unnecessary health care, and imports from China.

Sending people checks and hoping they will spend them wisely won't work any better than giving people treadmills and hoping they will exercise on them. They will probably just hang clothes on them instead. Some newer models even come with extra hangers.

Q: How did we get into a Methadone Economy and how do we get out of it?

For years, our economy grew because we all spent large sums of money justified by the illusion of housing and stock value appreciation—this was like readily available heroin. This spending created an abundance of stores, restaurants, hotels and malls that needed our money to stay afloat. One could say that consumers became addicted to a certain level of consumption based on the assumption of ever-rising home values, and businesses became addicted to a certain expectation of demand based on ever-rising consumer spending.

The heroin supply of housing stock appreciation suddenly dried up. Given the choice of sending the country cold-turkey into a depression or helping us to keep spending, the Government has chosen to keep spending, through annual deficits now projected to last until 2020. Like Methadone, this money is supplied by the Government and doesn't provide remotely the same high, but it does keep us from going into acute withdrawal. The aforementioned 2008 Stimulus Check was the first Methadone "fix." The 2009 stimulus package was the second. Even without a specific stimulus, future years will show large deficits both because of previous promises made, and because of –guess what – interest on the loans to finance the previous deficits. But if the government doesn't extend those deficits into the future – if it abruptly raises taxes or reduces spending — growth could once again dry up.

Therein lies the problem. Methadone is addictive too. And unless we find real, "clean" sources of growth not requiring the government to spend money, the Methadone Economy will last indefinitely, dependent on government spending to provide enough demand to power slow growth. True long-term economic rehabilitation requires a multistep process to wean the country off Methadone into an economy based on neither bubbles nor bailouts, but rather on sound, creative economic choices and policies. That is what *Why The Heck* is all about...and that's why we need your ideas.

If, on the other hand, we don't receive your ideas, we don't wean off Methadone, and the government becomes the dominant force in the economy, we will stagnate. Unemployment will remain chronically high. We will lose our position as the world's leader, and spend most of our time reveling in our past glories. Imagine Europe, but with slower trains.

Q: Why did you selflessly start this website and fund this prize? In other words, why are you such a great guy?

Why does anyone devote himself and potentially a large portion of his net worth selflessly to a cause, and then work tirelessly for the greater good, with no thought of personal reward?

Spite, obviously.

Here is where the compelling-love-and-betrayal part comes in. I was once engaged to an economics professor. She wanted to get married on a specific date. I couldn't get married on that exact date. Instead of switching wedding dates, her "outside the box" solution was to switch fiancés. Unbeknownst to me, I had just been her main fiancé, and she had an emergency backup fiancé for exactly this contingency.

The good news is that I walked away with a Lovely Parting Gift, which was the realization that I knew more about economics than she did. Her knowledge of other branches of economics, outside her own narrow area of specialization where she was quite justifiably well-respected, was on a par with that of your average golden retriever.[2] (Truth be told, he was a below-average golden retriever. If you shined a flashlight on a wall, he'd try to eat the light spot. Not once, not twice, but until the entire wall was covered with snout marks. We suspect he was strangled by his umbilical cord *in utero*.)

I mean no disrespect by this comment, not just because many economists who specialize in one area don't stay current in others, but also because my wife's golden, despite his brain damage, was an expert in behavioral economics. My wife tried to train the dog to stay out of the basement, by giving him a MilkBone to come back upstairs every time he went down there. Naturally he learned the opposite. Every time the basement door was even slightly ajar, he'd open it with his snout, and go sit downstairs until someone noticed, exactly like a behavioral economist would. At least one who likes MilkBones.

I also learned from my former fiancee that while economics has plenty of prizes, there is no prize for what seems to matter most—for inventing the best economic idea, period. So that's what I decided to offer. Yes, I created this prize out of a *soupcon* of spite, but mostly it was because I couldn't believe there was

---

2. Speaking of pets, the guy she ended up marrying owns four cats. In my experience, middle-aged guys who own cats fit into one or more of three categories: (1) They should come out of the closet; (2) They need psychiatric help; or (3) They want to kill James Bond.

no prize like this one already. Especially because America is a prize-crazed culture. There is a prize for almost everything, including for the person who looks the most like his or her dog. Almost everything, that is, except for the most important category of all: ideas to transition the Methadone Economy into one generating sustainable growth.

Ironically the fiancé's slight that generated this prize is now long since forgiven and (almost) forgotten (but not quite), so you might ask: "Why continue to offer the prize? Why continue to do something at great expense if the rationale for doing it no longer applies?"

Beats me, but as it turns out, most of government works on that exact same principle. Like the census. Why do we still count people in person one at a time, at great expense, every ten years? In 1790, you did indeed have to get on your Trusty Steed and go count people in person one at a time, at great expense. But that was because our forefathers lacked the seven basic tools needed to generate demographic reports—statistical sampling methodologies, computers, zip codes, social security numbers, MBAs, cubicles, break rooms, and lattes.

And why every ten years? It's because we use base ten, which in turn is because we have ten fingers. If we had eight fingers, like Homer Simpson (count 'em), we'd do a census every eight years. And don't get me started on Mickey Mouse (six fingers), Mordecai Brown (three fingers) or Colonel Sanders (finger-lickin' good).

Yet despite all that effort and all those fingers, the census isn't even particularly accurate. How can undocumented aliens be counted if they aren't documented? Do census takers say, "We're from the government and you need to fill out this official-looking form, but don't worry. Your secret's safe with us."?

And what constitutes a "person"? Do census takers who believe that life begins at conception ask women if they're pregnant? Or whether they are a week or two late and more than a little bit concerned? And why is a census necessary at all? With the previous census being six years earlier, how did anybody know the United States population reached 300,000,000 on October 16, 2006 at 7:46 AM? Not "sometime that morning," or "somewhere maybe around quarter of eight, but my watch could be a little slow and in any case I was in the break room because I had been working straight through since 7:15 so I mighta missed it" but 7:46?

My point is that I am not the only one who keeps doing stuff for no apparent reason. There is a whole chapter asking why we still mint pennies, for example, when the rationale for them has long since been forgotten, and when they are worth about a tenth as much as they were a century ago.

Q: Oh, come on. You can't be serious. No one writes an entire economics book out of spite.

That's not a question, but I'll answer it anyway.

I did indeed start the blog that morphed into *Why The Heck* out of spite for my previous fiancée. Likewise, the reason I decided to turn it into a book was also inspired by her. Even though she wasn't particularly bright, she had written an economics book. I figured, if she can write an economics book, how hard can it be?

But although I started this project out of spite, I don't want readers to think I am one of those permanently embittered characters who achieve satisfaction by lashing out at friends, colleagues and former benefactors, like Scrooge or Newman or Sarah Palin. Spite was only the reason I initiated the project. The reasons I've continued it to the point of publication are to (1) educate people on the state of the economy; (2) bring joy and laughter into the lives of others; (3) influence national policy in a very positive direction; and (4) meet women.

Q: What constitutes a great 'outside the box' idea?

First, the idea needs to be ORIGINAL. Someone recently posted: "Legalize prostitution and tax it." Interesting idea, and it would be original too — if I were collecting ideas on papyrus to present to the Pharaoh.

Second, the idea needs to be FOCUSED on ONE insight. Don't send in any "10-point plans to reform the tax code." To begin with, the likelihood that you have 10 original points in anything, especially tax policy, is nil. Also, nothing personal but no one wants to hear your stupid 10-point plan.

Third, compliance with the idea should be VOLUNTARY. Virtually every proposal described in *Why The Heck* requires nothing from anyone. No mandates, no making anything illegal, no required new taxes. *Why The Heck* philosophy is about aligning incentives and removing inefficiencies and hidden subsidies that distort consumer behavior.

Fourth, implementation of the idea could INCREASE TRANSPARENCY. This book contains two proposals that shine a light on various middlemen working directly or indirectly at taxpayer expense. Exposing their commissions to bright light will probably result in reducing them.

Where *Why The Heck* and you add value—and where you are most likely to win an award—is with logical but previously non-obvious innovations fitting some or all of those criteria. Remember, anything that makes the economy function and grow without government spending puts us one step closer to Methadone withdrawal.

Q: Are you really offering $1-million dollars for the best idea?

Yes, and though no one has won the big prize, I have already awarded two $100 prizes for ideas in this book, so try your luck/skill. All you need to "play" is the ability to think outside the box. You don't need a PhD. You don't even need a Masters. Or, for that matter, a BA, a high-school equivalency diploma, or even an up-to-date inspection sticker. Remember, economics professors don't have a monopoly on good ideas just because they have fancy degrees, tenure, and tweed jackets with elbow patches. Think of all the breakthroughs achieved by people with no credentials. Like in music, for example. Beethoven didn't have a PhD in musicology. Paul McCartney can't read music. Bob Dylan can't sing. And Britney Spears can't stay out of the tabloids.

Q: What are the best ideas so far?

Nice try, but if I told you now, you wouldn't have to buy the book, would you? The best way to find out is to buy a copy. You could just read *Why the Heck* in the bookstore in its entirety, ignoring the dirty looks from the staff, but you may find the ideas so provocative and inspirational that you'll want to take it home. Also, having the book handy makes it possible to study the rules to learn how to submit your own million-dollar idea.

Before you submit your idea and start looking at second homes, or even fill your tank with premium at a full-service station and pay with a credit card, I'll give you fair warning: There are some VERY good ideas in here with which you would be competing. As measured by TV, radio and print coverage, www.whytheheck.com is by far the most prominent economic idea-generation website anywhere, and as a result has attracted ideas that you won't find anywhere else but right inside the very book you are holding in your hands right now.

Q: What are the worst ideas so far?

"Worst" is in the eye of the beholder, of course, though sometimes the beholder has a darn good point. For instance, the *Guinness Book of World Records* once awarded the record for the "worst film editing job" to a movie theater in South Korea that, having decided that *The Sound of Music* was too long, edited out the songs. Likewise, a couple of ideas submitted to *Why The Heck* would stand out as the "worst" to anyone, even to that theater owner.

For instance, any idea beginning with "The government should subsidize" is probably a really bad idea. After all, consider some examples of what the Government actually does subsidize, directly or indirectly: mortgages, health care, gasoline, food. See a pattern here? We have become a country of oversized people driving oversized cars from oversized houses to oversized medical facilities. Even those instances where a subsidy is a good idea would not belong in this book, which in the broadest sense is about getting the most out of existing resources. The government does not need our help to spend money. Those guys do a fine job of that on their own. They've certainly had plenty of practice.

Besides spending (our) money and causing overconsumption, the other problem with subsidies is that they are not market-driven. You may be shocked, *shocked* to learn that sometimes people get subsidies for political reasons. In other words, the government rather than the private sector allocates resources, comrade.

Next, any idea with the word "Syracuse" in it is probably also not a good idea. Nothing personal against Syracuse (my mother's family lived there) but—aside from a few good college hoops teams and some air conditioners—I am not sure what that city has produced since the Erie Canal closed. We're talking about a city where you can't just hail a cab. You have to call one. Maybe that's not so unusual. But at the airport?

And given that no World's Fair has attracted enough visitors to break even since the Johnson Administration, another World's Fair probably isn't such a good idea either.

Therefore an excellent candidate for the "worst idea" would be one I received in February 2009: "The government should subsidize a World's Fair in Syracuse."[3]

More typically, though, a bad idea sounds really good on the surface, until you dig into it. For instance, in the depths of the economic crisis in early 2009, someone suggested creating a new currency and sending $1000 worth to every taxpayer. This new currency, called "Stimulus Stamps," would have value for only 90 days. At that point it would expire. That meant it wouldn't add to the deficit or create hyperinflation. But you'd have to spend this currency right away, because you'd lose it all

---

3. I asked the submitter of this idea why anyone in their right mind would ever want to attend a World's Fair in Syracuse. In a rather exasperated tone implying that you'd have to be an idiot to even ask, he replied: "Obviously, because the Government would be subsidizing it."

if you were still holding onto it on Day 90. Likewise, whoever you spent it with would want to quickly spend it again. Thus the "velocity" of the currency would be very high, stimulating the economy with every purchase, but without any inflationary effect on the money supply because the currency would have no lasting value. His observations are all well-and-good if you're the one spending the Stimulus Stamps. But what if you are the one who's receiving the Stamps for a $1000 purchase? Is your reaction: "Oh, terrific. I'd way rather have $1000 in Stimulus Stamps that are going to expire than in stupid old currency backed by the federal government"?

No, your reaction is: "Get out of my store before I call the police." Oh, sure, maybe on Day One you'd accept these Stamps, but not for face value. The value would have to be negotiated. Each day, for each storeowner, the rate would be different. So, if each transaction has to be negotiated, why bother with currency at all? The whole point of currency — that it represents a fixed known accepted value — would apparently no longer matter, why not just replace all our current currency with Stimulus Stamps and barter with it?

Q: What has been the reaction of the actual people in Washington?

If by "actual people in Washington" you mean, actual people who are like you and me, but who happen to live in the District of Columbia, it's been great. Everyone who reads or hears these ideas likes most of them. However, if by "people in Washington" you mean anyone whose opinion matters even a little bit, the answer is that they get their advice from economists. Is the Government getting its advice from the right people? One economics graduate student posted the following on the *Why The Heck* blog:

> I spend most of my time studying, teaching, or doing original research into a topic. But the topics are too narrow to ever change the world. The idea of ideas is actually pretty foreign to most economists.

Economists are trained in, well, economics. So they apply what they've learned. Keynesian economics that was used to justify ongoing deficits was invented maybe 80 years ago. Is anything else still done the way it was done 80 years ago, other than Major League Baseball? Maybe it's time for a breakthrough to a new generation of economics — call it "heckonomics" — that does not rely on spending money. Recall that Keynes was "outside the box" in his era too, and if President Roosevelt had stuck with Hoover's inside-the-box idea of balancing the budget during hard times, the Depression might have been worse. There is no reason for the Government to discard outside-of-the-box ideas today, either. They might be discarding the next Keynesian-level breakthrough.

Here's another way of looking at it. More than 300,000,000 people live in this country. Of those 300,000,000+, maybe 1000 have the actual day job of coming up with ideas to solve our economic problems. It's not like they've solved anything by any means other than throwing money and regulations at the problems, so who's to say that we shouldn't give the remaining 99.999% of the population a chance to contribute their ideas? That's what *Why The Heck* has done, reviewed ideas from anyone willing to contribute them rather than from the people who do this for their day jobs already but who have collectively let us down, a la Neville Chamberlain, the Chicago Cubs, or Puxatawney Phil.

And that's why *Why The Heck* is needed. It's the people's economics, not economists' economics. As Einstein said, "A problem cannot be solved by the thinking that created it." It is up to you and me to find the nuggets in this book and post more on www.whytheheck.com and get the ideas to the politicians who make the rules. So mail these solutions right to your Congressperson. Even if it means you have to buy an extra copy of this book to do so.

Q: Are you looking for Democratic-type ideas or Republican-type ideas, neither or both?

There is no pre-existing political agenda. The Green Dividend is designed to painlessly raise the price of gasoline and hence could be considered a Democratic idea because such a price increase might be viewed as a gasoline tax, and that is something Democrats support (at least according to Republicans). However, the *Why The Heck* insight about gasoline economics is that the price needs to cover the cost of defending our oil supply lines from threats real and imagined. Cost Accounting 101 calls for allocating resource costs to the products incurring them. If our oil were all produced domestically, our defense budget could be quite a bit less than it is. Being so dependent on imported oil carries costs well beyond the price of the oil itself. Those costs should be part of the price.

Just by citing basic accounting, the argument that gasoline is priced too low can be made without a single reference to global warming and carbon dioxide emissions. Unless Republicans don't believe in accounting or unless they think our defense budget is high because we need to defend ourselves against KAOS, they should agree with this too.

On the other hand, because most of the solutions in this book call for tearing down barriers, making markets more efficient and transparent, and enhancing voluntary choice, many solutions should appeal to Libertarians, most of whom are Republicans.

Or neither party, as in Chapter Eleven on payday loan reform. *Why The Heck* philosophy supports the Democrats' view that a 360% Annual Percentage Rate (APR) on payday loans is much too high, but then offers a solution that would probably reduce the APR by two-thirds through free market reforms and minimal, temporary government involvement, without the price controls favored by the Democrats.

Sure, there will be room for political debate in these proposals. Consider Chapter Four's ShopAmerica Gift Card again. It's voluntary and free and would encourage more spending, meaning that there are no trade-offs, no grounds to oppose it. The Republicans will argue that the ShopAmerica Gift Card is a reason to send out more Stimulus Checks instead of spending more on government spending programs. So if John McCain had been elected, we might have been approaching a stimulus plan very differently. But he lost. Still, give the man some credit. He managed to almost carry Indiana, a feat no Republican presidential candidate had pulled off since 2004.

The Democrats would respectfully disagree with McCain and the Republicans and say, no, let's just apply the ShopAmerica Gift Card to government payments already being made.

Respectful disagreement is not so bad. That is exactly the job of lawmakers, as foreseen by our Founding Fathers when they created the world's first constitution-based government. They foresaw how honorable men (and now women) who have sacrificed their careers to serve the public could start with differing opinions, but then—through respectful debate, compromise, committee testimony, checks-and-balances, and, most importantly, Listening to the Will of the People—keep working out those differences until Congress' approval ratings drop to zero.

Your job on my blog and my job in this book are to think big, think Outside This Two-Party Box, and propose ideas which, like the ones you are about to read, are too promising not to implement, no matter which party is in charge.

Part One

Ideas Requiring One Lousy Little Sentence to Implement

Two economists are walking down the street and one spots a $10 bill on the sidewalk. He says to the other: "Look—there's a $10 bill on the sidewalk." The other, without looking down, replies: "Nah, if there were a $10 bill on the sidewalk, someone would have picked it up already."

Likewise, you would think that if indeed there were ideas to improve the functioning of the economy, even in minor ways, that would require only one sentence to implement and that would be totally voluntary besides, the Government would have picked them up already, right? Read these first three chapters before you answer that question.

Chapter One proposes a simple one-line Executive Order that would allow businesses to forego handling pennies if they agree to "round down" and never up, generating savings for everyone involved because it costs most businesses more than a penny to handle a penny.

Chapter Two proposes a one-line addition to the current Form 1040 that would dramatically facilitate tax refunds and save everyone money...including the Government itself.

Chapter Three proposes a one-line internal government directive that will spur hotels to become more "green," while reducing government spending and room prices.

True, these minor ideas would generate only minor economic improvements, but Chapter One also proposes a theory called Broken Stimulus. It is based on the Broken Windows theory of crime (which states that police actions to prosecute small transgressions makes major crimes less likely) and states that people will trust the Government more on major economic issues, if it can make minor economic changes that inspire confidence. When people trust the Government to create a better future, they will spend their own money more freely. Therefore minor, confidence-building economic improvements could assume a level of importance way beyond the actual dollars involved.

You can already see the importance of restoring confidence in government — consumer confidence rose considerably from its low point way before spending rose enough to make unemployment start to decline.

You'd have to acknowledge that President Obama—whether you agree with him or not—inspires much more confidence generally then either of his two predecessors because this administration is not remotely as entertaining as theirs were. No Travelgates, DNA tests, or declaring Mission Accomplished in wars which had barely begun. Obama never gets misunderestimated. So far no one in this administration has shot any old men in the face. Finally, even though Obama and Biden are still in their first terms and presumably are still getting to know their jobs, rarely is the question asked, "Is our leaders learning?"

CHAPTER ONE

Pennies from Heck: How America is Being Crucified on a Cross of Zinc

In the immortal words of the great philosopher Julie Andrews, let's start at the very beginning, a very good place to start. The first proposal in this book—and the first proposal featured on www.whytheheck.com to gain any attention outside of my immediate family—has been a proposal to phase out pennies, voluntarily.

The penny is one of those things, that if it hadn't already been invented, no one would propose coming up with it, and yet now we can't get rid of it. Sort of like the Electoral College or pit bulls or Canadian football or Al Sharpton. If pennies didn't exist, I doubt very much that anyone in Congress would say: "Let's create a coin with one-fifth the value of a nickel, that inconveniences consumers, banks, and merchants, that most people lose anyway, and that pollutes the environment."

Yet pennies are still here, because it's always hard to make change happen (not only is no pun intended, but I didn't even notice it myself until now). For instance, despite the overwhelming lack of evidence for their efficacy, people spent centuries worshiping the sun, applying leeches and sacrificing virgins. It even took six years to convince the inventor of basketball that the basket should have a hole in the bottom.

Point well taken, right? But even so you might ask: "With all the problems the country faces, why on earth would you waste time dealing with something as trivial as pennies?"

That is a very good question. Pennies are trivial indeed. But please suspend your judgment until you've actually read the proposal because the reason for the proposal's importance far transcends the penny's monetary triviality.

Consider the aforementioned "Broken Windows" crime theory.[4] The level of violent crime in the New York City subway was dramatically reduced when the authorities aggressively addressed trivial transgressions like graffiti and turnstile-jumping.

Likewise, perhaps our collective lack of confidence in the Government's ability to get the big things right is the inevitable result of its habitual inability to solve small and persistently annoying everyday problems. When you fear the Government

---

4. I don't normally cite actual books in these footnotes because formal citations require remembering whether the author precedes the title, whether you put the author's first name before the last name, where the commas go, what information you put in parentheses, italics etc. Not to mention digging up the identity of the publisher, like anyone cares who published the thing. When was the last time someone recommended a book to you by saying, "One of the best-published books I've ever read"? But because I didn't come up with this Broken Windows theory, I'm afraid my editor would insist on my crediting the actual folks who did, so here goes: George Kelling and Catherine Coles, *Fixing Broken Windows: Restoring Order and Reducing Crime in Our Communities* (Free Press, 1996).

cannot solve big problems because it can't seem to even solve small ones, you may become reluctant to spend your money because you expect hard economic times ahead. Call it the Broken Stimulus theory.

For instance, in my town of Wellesley, Massachusetts, there is a red light hidden from view over the crest of a hill on a road where cars routinely drive as though they expect no traffic lights, and certainly not a traffic light hidden from view. Consequently, there were many accidents and near misses at that intersection. So the state, wisely at least in theory, put up a sign before the top of the hill to indicate whether the out-of-sight light ahead is red or green.

What a great idea, I thought, and such an elegantly simple solution. Ah, but even when there shouldn't be any details, the Devil manages to find his way into them. How hard is it to put up a sign indicating a red light up ahead?

Too hard for the Commonwealth of Massachusetts, it seems. In practice, the sign does exactly what I said: It indicates whether the upcoming intersection's light *is* red at the particular moment at which you are passing the sign. It does not provide what would be the much more useful information of indicating whether the light *will be* red once one drives over the hill to encounter it. The distance between the sign and the light is great enough that the light changes about 20% of the time after a driver passes the sign but before reaching the light.

Worse than being simply ineffective 20% of the time, the sign establishes a false expectation by incorrectly indicating 20% of the time that the light will be green, and therefore could cause even more accidents than no sign at all. Every time I pass that sign, I think, "Why did I vote for these halfwits?" Naturally my tendency is to automatically oppose much larger state programs as well, on the theory that if they can't get a simple sign right, how can I trust them to spend a billion dollars wisely? For just that reason, the first set of *Why The Heck* proposals suggests small but creative solutions. These proposals send a simple message to the Government: Implement creative win-win solutions for the visible everyday annoyances like pennies, and people will give you the benefit of the doubt when you undertake initiatives on a grand scale. And what could possibly be a more visible everyday annoyance than pennies?

How annoying are pennies? Some people might say they "like" them, but the simple fact is that most people don't. The population grows by about 30 million people a decade. In the last decade the Government minted more than 60 *billion* pennies. So the number of pennies theoretically in circulation grows *2000 times faster* than the population.

Obviously the Government wouldn't need to mint nearly that many pennies if people actually reused the ones they get in change. Instead, most of these pennies find their way into coffee cans, sofas, and the occasional toddler. That's 2000 pennies, or $20, lost—a *de facto* tax, but, uniquely among taxes, one that raises no revenue. Pennies are also a waste of zinc (copper is too expensive to use), much of which is imported. They also cost so much to mint that the Government loses about half a cent on each penny, or another $15 per person.

Taken together, every single person who is born into or who steps into this country starts out $35 in the hole, simply because the Government has decided to import large quantities of zinc, tax us each $15 to mint this zinc into trash, and then tax us again, $20 apiece, to dispose of it.

And while we're on the subject of disposing, the zinc we don't import comes from Alaska, and Alaskan zinc mines dispose of their tailings in some of the same rivers where salmon spawn. Pacific salmon are disappearing at an alarming rate for many reasons, river pollution among them.

There is simply no argument for keeping them (meaning the pennies). The zinc lobby ("Americans for Common Cents"), and a few consumer advocates with too much time on their hands whine that eliminating the penny would allow businesses to "round up" transactions to the nearest nickel, thus costing consumers money. That argument has stalled every piece of legislation on the topic ever introduced.

And this is where the creative win-win *Why The Heck* solution comes in.

Now apply the Broken Stimulus theory to government and to this problem in particular. If, in fact, there was a way to solve the penny problem without tying up Congress, without a heavy-handed mandate, and without consumer or business opposition, then confidence in government would probably increase. (Broken Windows would say that solving one small problem is insufficient to increase order. The same is probably true with Broken Stimulus and increased confidence, so it's a good thing that there are two more such ideas in Part One of *Why The Heck* alone.)

Fortunately, there is such a solution. President Obama could sign a simple one-line Executive Order saying, "Businesses that agree to round down their cash transactions may refuse to handle pennies." It is well established that the Executive Branch can make exceptions to the Legal Tender rule (such as allowing toll booths to refuse pennies) without involving Congress. That means that are one signature away from ending the tyranny of the zinc lobby, one signature away from solving the most visible everyday examples of government paralysis in the face of special interests, and one signature away from starting people on a path to believing in government again.

Why would this solution work if compliance is completely voluntary? Look at it in theory first. In addition to costing more than a penny to mint, time-and-motion studies indicate that a penny costs more than a penny for most retailers to handle. Businesses can thus save themselves money by rounding down instead of counting out pennies in change. It's counterintuitive but true: Giving out more money in change, in the form of nickels and dimes, costs less than giving out less. Most consumers will appreciate the extra savings and faster checkout lines. Everybody wins.

Any businessman who disagrees with the savings arithmetic is welcome to keep handling pennies. Customers who prefer to pay with pennies may continue to frequent those businesses. No one is being forced to turn their pennies in. In fact, no one is being forced to do anything, just permitted to.

Now let's look at in practice. When the *Why The Heck* blog first proposed this idea in January 2009, a reader brought it to a bakery in Concord, MA. The bakery, located across the street from a commuter rail station, had a peak-load problem during the morning commuting hours. People would often abandon the slow-moving checkout lines to catch their trains. The owner contacted me to offer support for my proposal, which she felt would speed up the lines. I suggested that, rather than wait for my proposal to be enacted, she take matters into her own hands by simply doing it on her own. Specifically, I suggested that she simply round down cash transactions to the nearest nickel and refuse to take pennies in payment, technically a violation of the law.

The significance of Concord being the birthplace of the first revolution—and that we were engaged in civil disobedience in the town where civil disobedience was invented[45]—was not lost on the local press. We received so much local media coverage that almost immediately about 30 more businesses in Concord signed on to the protest. Most of them are staying with it even in this sagging economy and many are reporting an increase in sales and even tips as customers show their appreciation for no pennies.

The extra pennies aren't being wasted. People are encouraged, in a very Lincolnesque way with the signage downloadable from the *Why The Heck* website, to put their pennies in a charity jar. Even though tips and sales are up, charitable contributions have risen as well. It appears that the average customer, out of gratitude for not having to deal with pennies, is actually leaving more money behind at the bakery now even though they are getting more money in change.

The Concord Teacakes bakery that started it all is so convinced that this is an idea whose time has come, that they hung a plaque to mark the location for future generations. This being Concord, it was inspired by Emerson's famous "Shot Heard 'Round the World" quatrain:

> *To the rude lobbyists who corrupted our good*
> *Congressmen to support their zinc*
> *Here the embattled merchants stood*
> *And said, "We think your pennies stink."*

---

5. For those of you who were absent the day this was covered in your eighth-grade history class, Concord's Henry David Thoreau invented civil disobedience by refusing to pay his poll tax to support the Mexican War, and landed in jail as a result.

So, both in theory and practice, voluntarily phasing out pennies is an idea whose time has come. More importantly, don't overlook the impact of the "Broken Stimulus effect." The significance of this easily implemented, creative, win-win one-line Executive Order far transcends the penny. It signals a willingness of the Government to "Think Outside the box" and find innovative solutions for many of our more pressing national problems as well. Consumer confidence will climb with each rounded-down money-saving cash transaction, transactions which—instead of being everyday reminders of government incompetence—become everyday reminders of government's ability to solve problems.

Perhaps the penny isn't worthless after all.

COMMENTS

As with every proposal in this book, we provide some reaction as well. Because many *Why The Heck* proposals sound too good to be true, we always invite comments on the blog and, in this case, also tallied comments on the many other blogs that picked up this particular idea.

The comments were mixed. While the supporters "got it," there was a general sense among the detractors that, even though the Executive Order would allow businesses to refuse pennies only if they round down, businesses would nonetheless figure out a way to secretly raise prices—as though one could keep price hikes a secret.

POSITIVE COMMENTS

"If I just put the pennies I've accumulated back into circulation, the country wouldn't need to mint any more," would be typical of the comments that found the observations about the growth of the penny supply compelling.

Another person asked: "Why do seafood restaurants take pennies? Salmon are disappearing from our rivers, and when those restaurants handle pennies, they actively contribute to the fish's disappearance by encouraging the zinc companies to mine more and pollute more.

On a more practical level, several people noted that, without pennies, there would be more room in the cash register for dollar coins, which every other civilized country adopted decades ago.

Another person asked, "Why put pennies in circulation in the first place? If the Government insists on minting them despite all economic arguments to the contrary (and despite this new idea), why not truck them directly from the mint to the landfill and eliminate the need for us to be the ones responsible for throwing them out?"

Yet another person asked the Broken Stimulus question without phrasing it as such: "How can we trust the Government to spend $800 billion if they can't do this?"

A few business owners wanted to know how they could take part in the protest and refuse pennies. Well, refusing to handle legal tender counts as an Official Serious Violation of Federal Law, way more serious than tearing tags off mattresses. Therefore, I would say, boys and girls, don't try this one at home. That would really be shortsighted. You could end up being ARRESTED and be put on a CHAIN GANG, like PAUL NEWMAN or WOODY ALLEN, and/or have to SHOWER in front of the OTHER INMATES. No, trying this at home would be very shortsighted indeed. Instead, try it at work.

That's where your actual customers are. You can download the signage to support this protest from www.whytheheck.com.

Finally, another reader found three reasons to get rid of pennies that no one else had thought of:

- Toddlers choke on them (clearly a public health problem of epidemic proportions that the zinc lobby has shrewdly kept the media from making us aware of);

- The nation's Strategic Zinc Reserve should be allocated to where it is critically needed, such as sore-throat remedies; and

- There is nothing more frustrating than trying to guess how many are in the jar.

She concludes: "On the tree of currency, the penny is a dead branch waiting to be lopped off."

NEGATIVE COMMENTS

Why The Heck invites negative comments. Many such comments have been used to refine these ideas prior to publication. In addition to those refinements, there have been instances — notably the infamous Stimulus Stamps — where I thought something was a good idea until the comments were posted. Yes, it's true. It took an onslaught of comments for me to remove Stimulus Stamps from the Idea of the Month heading. (Don't tell anybody.) As a result, the prize rules encourage and reward comments of the negative as well as the positive variety.

So I mean no disrespect whatsoever when I observe that the negative comments on this particular posting came mostly from people whom one would have assumed Darwin would have taken care of years ago.

First, some people confused "rounding down the cost of total transactions" with "rounding down prices and other things." One person, for example, thought that his state would never go along with reducing its 6% sales tax to 5% just to save pennies.

Second, people confused "rounding down" with "rounding up," one person complaining that his Slim Jim, which had cost $1.06, would now cost $1.10. I was heartened to see that the opposition was eating Slim Jims. Darwin was hard at work.

Third, people confused "commenting on this penny proposal" with "commenting on whatever else they feel like ranting about," and complained that getting rid of pennies would lead to hyperinflation, bankruptcy and Godlessness. OK, I just made up the Godlessness part, but admit you believed it for a second.[6]

And speaking of Godlessness, someone asked what Jesus has to do with pennies. After a little head-scratching I realized he was referring to the chapter title: "America is Being Crucified on a Cross of Zinc." That's actually a take-off of Williams Jennings Bryan's famous mantra: "America is being crucified on a cross of gold." Readers unfamiliar with the quote are no doubt unfamiliar with Bryan too, so here is a primer. Bryan ran for president unsuccessfully a record three times, and then capped off his career by representing the creationists in

---

6. I spoke too soon. In preparation for this book, I re-checked Michael Barone's old *U.S. News* blog entry on this subject for the first time in months. Along with the 400+ mostly positive secular comments in total, he has indeed received more than 20 comments on this proposal in which God was cited as a reason not to get rid of pennies. The irony is that www.KruseKronicle.com ("Commenting on Ministry and Culture in the Twilight of Western Christiandom"), which fears God more in an average day than most of us do all year, long since came out in favor of this proposal.

the Scopes Monkey Trial. For more information, google on "William Jennings Bryan," "Populists," "Boy Orator of the Platte," "wrong side of history," and "poor script selection."

One more negative comment offers a thoughtful and valuable caution for President Obama if he were to announce this Executive Order: Keep it in the context of what we are now calling the Broken Stimulus theory. Don't pretend you are solving the world's problems by signing this one-line order. Either sign it in a group of other Executive Orders as part of a package of Broken Stimulus fixes, or else make very clear that this is simply a symbolic act to show that you are thinking outside the box. It would backfire if you called a press conference and announced this rounding-down option with great fanfare and expected people to react to the substance. Whenever you hold a press conference, expectations are high. So maybe make this an "Oh. By the way," type of Executive Order and downplay it, like presidents usually do at the end of their terms when they pardon prisoners who have contributed to their campaigns. Or maybe just have someone at Treasury announce it.

The serious lesson to be drawn from the negative comments: Even a simple one-line Executive Order creating a business opportunity for which compliance is voluntary will incite a negative reaction from some quarters just because the Government is doing it, and those guys always screw up everything. Therefore, this initiative will need to be supported by some consumer education and the Government will need to follow up and fix some other Broken Stimulus problems before people believe they can trust the Government again to spend money wisely.

CHAPTER TWO

Reducing the Cost of Being Poor

The Broken Stimulus theory of government can be applied to many other opportunities, not just pennies. Recall that this first set of chapters is about proposals that can be executed by the Executive Branch alone, without even breaking a sweat or killing a tree or, if they execute these proposals online, maiming an electron.[7] Remember what the Broken Stimulus theory tells the Government: Show the people some small and creative innovations, and you will earn their confidence that you can handle big problems as well.

Those of you who can afford to buy this book probably don't know how much it costs simply to be poor. Start with something like direct deposit of a paycheck. You just take that for granted. But what about people who don't have a bank account at all? They have to go to one of those check-cashing storefronts you've probably only seen in the background in some Martin Scorcese film, or else maybe when you took a wrong turn somewhere,

---

7. Just for the record I do indeed know that posting things online does not maim electrons, injure them, or even stress them out much. Indeed with proper care and nutrition, they can live to be a hundred.

like Sherman McCoy. A later chapter will address the specific and somewhat more complicated issue of getting people bank accounts, but this first part of the book is about ridiculously easy fixes.

One such fix has to do with tax refunds. Many working-class people get their taxes prepared by a tax preparer. No, not a fancy CPA like yours, with a fancy marble conference table with a fancy bowl of those fancy little fruit candies where you suck on the hard outside at first and then there is a gooey part in the middle bursting with natural and artificial flavors that you can chew on and hope that your fillings don't get sucked out, or if they do then you hope that your accountant will say it's tax-deductible because you were meeting with him at the time. If your accountant doesn't seem to have any left, it's because I ate all of them (meaning the candies). No, these are storefronts in "transitional" neighborhoods temporarily rented and outfitted with a few computers and a few tax preparers. Their clients usually have pretty simple tax returns but lack the computational skills to do their taxes themselves.

How many of these preparers are there? Enough so that H&R Block is the largest seasonal employer of white-collar workers in the country, a fact known apparently to almost every *Jeopardy* fan and player except Ken Jennings, whose record-setting reign ended by not knowing this tidbit. (I myself would have guessed "Hallmark" and I also would probably have forgotten to put it in the form of a question, so I would have ended up owing Alex Trebek money.)

As a reader you are probably thinking, "Oh, he is about to advocate for simplifying the income tax system, to make it fairer so that people can file without help." No, I'm not. That may be a good idea too in theory, though 'simple" and "fair" often prove in practice to be opposites of each other. Remember that the first part of this book is about quick fixes to the Broken Stimulus problem to restore confidence, and tax reform would hardly qualify as a quick fix.

No, the opportunity I see takes place after the taxes have been prepared and before they get filed. The tax preparer says, for example, "It looks like you are entitled to a refund of $500." So far, so good, right? But then he often says something like, "You might have to wait a couple of weeks for that refund. If you prefer, we can give you a check for $475 right now."

Put yourself in the place of that client. Almost by definition, he or she is living paycheck-to-paycheck, since that is the demographic for commercial non-CPA tax preparation. This client has bills to pay, possibly credit card bills that themselves carry a very high interest rate. The enticement of the immediate cash is great, and roughly 8,000,000 people a year agree to the deal and take the money. After all, what's $25 off the top if you can take $475 home?

$25 is 5% of $500, that's what it is. And if you were going to get your $500 refund in two weeks anyway, it's an implied interest rate of almost 3% *a week*.

How can tax preparers get away with that? Because the client is there, the client owes someone money, and the client can't do the math. It is not a free and functioning marketplace because one side knows what they are doing and the other side doesn't.

Now, you're thinking that Al is going to propose making those "tax refund anticipation loans" illegal. No. If there is anything the country should have learned from Prohibition, it's that markets cannot be legislated out of existence. *Why The Heck* proposes something much simpler and more creative: Add a line to the Form 1040 and 1040A, right after where it tells you the size of the refund due to you, that says: "Check here if you would like 99% of this amount deposited to your bank account immediately upon electronic receipt of this return, in lieu of your full refund being mailed to you." Those without bank accounts could be offered a check that would print out immediately in the tax preparer's office, perhaps at 98%, so the tax preparer is compensated 1% for use of their time and printer.

Instead of getting $475, the client would get $495 (99% of $100). Government would actually make money on this transaction because, assuming the IRS would have sent the $500 in two weeks anyway, the interest charged is still a multiple higher than the Treasury Department's cost of short-term borrowing.

The financial term for this is "disintermediation." It's cutting the middlemen out of the process, using a combination of the Internet, the world's lowest cost of capital, and a pre-existing "loan" document (the Form 1040). Everybody benefits except the middlemen. One of the recurring themes of proposals in *Why The Heck* is middleman-elimination. Ironically, I myself am a middleman in my "day job." To find out where, you have to keep reading.

COMMENTS

The comments on this proposal were almost all positive. One individual pointed out that two major tax preparers had recently settled a massive lawsuit for misrepresenting these loans. "It's bad enough that they make them. Do they have to misrepresent them too? That's like the Woody Allen line about the two old ladies in the Catskills. One says to the other, 'You know the food here is terrible.' The other replies, 'Yeah, and the portions are so small.' "

Someone else added that Jackson Hewitt, another tax preparer, had been allegedly urging people to overstate expenses so that Jackson Hewitt could make larger loans – until the Justice Department filed suit. Obviously, if you take tax preparers out of the loan business, you take away the incentive to maximize refunds in order to make large loans.

NEGATIVE COMMENTS

Some of the negative comments were of the "this-is-too-good-to-be-true" variety. "There MUST be a reason we aren't already doing this or half of the other ideas on this site."

Someone suggested the reason could be that the government is checking for fraud before sending out the refunds ten days later. But all they are doing is checking eligibility and math. It takes much longer than ten days to determine fraud. The government basically does nothing in those ten days that they can't do electronically in an hour.

Another comment in its entirety: "I am finding myself getting more and more upset as I read these things. These postings seem so easy to do. The current administration just wants to spend and spend but why aren't they looking for solutions that don't require spending? Where are the Republicans when we need them? I feel like I don't have any representation in the Government. Everyone just wants to spend more than everyone else."

I wondered, was this guy talking about those same Republicans who brought us the Stimulus Checks? The tax cuts for the wealthy when the budget was already in deficit? The proposal to privatize Social Security so we could invest our Social Security account in the stock market? Or perhaps there is another Republican Party out there that does only smart things, that the zinc lobby has shrewdly kept the media from making us aware of? Those zinc bosses are so clever...Joe Hill wouldn't have stood a chance.

CHAPTER THREE

Behind the Green Door

Everybody knows that "going green" is expensive. Case in point: If you google on the words "Reducing," "carbon dioxide," and "Emissions," along with the word "Expensive" you get 1,060,000 hits.

But if you do the exact opposite search and Google on "Reducing," "carbon dioxide," and "Emissions," but this time along with the word "Inexpensive,"   the number of hits falls to 312,000, *which is 848,000 less hits, or 71%*!

Scientifically that analysis leads to the inevitable conclusion that we have to reduce emissions by 71% within the next six months in order to keep the earth from crashing into the sun.

You're not convinced? You must be one of those troglodytic global-warming deniers Al Gore warned us about, who refuse to accept the seemingly incontrovertible proof of climate change, even though I just spelled it out for you. You would say, "This analysis is fascinating but I think there might be some flaws in your methodology and conclusion. Specifically, it makes no sense whatsoever."

And you'd be right, of course. The analysis is indeed completely wrong on two accounts. First, when you take away 312,000 hits from 1,060,000 hits, you get 748,000 less hits, not 848,000 less hits. Second, it should be "fewer" hits.

You know what else makes no sense? The idea that somehow reducing energy use costs money. Look around your very own home, for example. If you turn down the air conditioning to save money, you spend less money, not more. And you know where people waste even more resources than at home? In hotels, that's where. Wasted resources go beyond energy, so it's not just that guests leave the air conditioning on all the time. It's that chambermaids are constantly replacing those little bottles of goopy fluids, scrubbing your bathrooms, vacuuming your floors, and (though less now than historically) changing your sheets. Since hotels don't charge for any of that stuff separately, people use — and waste — much more than they would if they were paying for it separately. That's why you don't see tiny bottles of shampoo being sold in stores, or people hiring maids to scrub their bathrooms every day.

Why the Heck proposes that hotels offer no-frills "Green Rooms." As with most other ideas in this book, the proposal is not to mandate that the hotel industry replace all rooms with Green Rooms, but rather just to offer them as an option.

Here's how the Green Rooms option would work. First, instead of those little individual soaps and disposable bottles of shampoo, conditioner and body lotion (who even uses their body lotion?), Green Rooms would have a dispenser for these fluids.

Consider your office as an example. I realize a few of you don't have offices because you still have one of those rare jobs in manufacturing that hasn't yet been offshored, but what I am about to say about offices is probably also true on the floor of your plant as well.[8] In your office, they no doubt have a men's room. You would probably be surprised if I tell you that your men's room is a best-practice benchmark for something, but it is. Specifically, it conserves resources. It doesn't offer little bars of individually wrapped soaps over the sink that everybody tears open, uses once and then disposes of, at which point an immigrant wearing a uniform comes in to replace them with brand new individually wrapped bars of soap for the next guy to tear open, use once and then dispose of.

No, of course not. That would be a ridiculous waste of resources. Instead, your men's room has one single, highly economical, dispenser of soap which saves resources because most guys don't use it at all.

So why can't they offer the same in hotel rooms? Why isn't there a big dispenser of various fluids over the sink and another one in the shower, just like in some health clubs? Those small bottles and bars cost exponentially more than large dispensers. A Green Room would do exactly that.

Second, in a Green Room, if you want changes of sheets and towels you'd pay extra. In fact, maid service would be an extra *a la carte* request, rather than hidden in the room rate. Who among us scrubs their bathroom every day when they are home? And are we really dropping so much stuff on the floors of these rooms that they need to be vacuumed every day?

---

8. Also, it makes me feel like a Regular Guy to say "floor of your plant," like I've ever actually visited one.

Third, hotels which are now being designed or renovated could also build in a meter and charge separately for air conditioning, thus encouraging people to use the air conditioning only when they need it rather than leave it on the whole time. As with anything else, if people are paying for it, they will use less.

These three simple steps would allow about 20% lower room rates for those Green Rooms. Lower room rates will boost occupancy. Why on earth the hotel industry hasn't figured this out yet for itself is anyone's guess. And that's where the federal government comes in. A simple one-line internal directive saying "Federal employees must stay in Green Rooms if available and if Green Rooms cost at least X% less," would immediately spur the hotel industry to adopt this option, and save the Government and taxpayers some money. Note that the CO_2 emission reduction argument is moot in this proposal. It doesn't matter whether you believe in Global Warming or not. Reducing resource use saves money, period.

Maybe government employees might object a little bit, but, hey, if it meant retiring on a full pension after 30 years I would happily make my own bed a couple of times.

Once again, though I suspect this idea would catch on in short order and people would start demanding these rooms on their own, this particular proposal, like the other two preceding it, would not save much actual money. Rather it would set an example that resource use reduction can save money, and it would encourage people and policymakers to start looking for other examples.

This is also yet one more case for the "Broken Stimulus" theory. It creates a win-win innovation, that people notice and appreciate enough to increase their confidence in government.

COMMENTS

Comments on this posting were uniformly positive except one. There were many reasons people thought this would be a good idea.

POSITIVE COMMENTS

First, someone contrasted the "bundled" pricing which hotels use to the *a la carte* pricing of the airlines, which these days charge for everything separately.[9] And I do mean "everything." One European airline, RyanAir, even floated the idea of charging to use the lavatories. The difference, it was noted, was that *Why The Heck* is proposing a la carte Green Room pricing as an option, not as a requirement. Tough to argue with allowing businesses to offer options.

Second, several people noted that no one actually has any clue what it costs to clean a room, to run an air conditioner, or even what those little bottles of shampoo cost, but that any time you don't charge extra for extras, those extras will be overconsumed.

Someone else noted that in Japan, they already do this, at least as far as energy use is concerned, and there is nothing optional about it. When you leave the room all the power goes off. Then

---

9. Recently I needed to board a plane but had forgotten my ID. I was sure the airline was going to send me home, but instead they couldn't have been nicer about it. In those situations their policy is simply that when you go through security you undergo a thorough search, just like the people who are "randomly" selected. Then you can board the plane just like everyone else. Needless to say I was quite grateful to be allowed to get on the plane at all, and totally understood the need for the extra inconvenience. The irony is, here is the ONE THING the airlines offer which I would HAPPILY have paid $100 extra for, recognizing that it was my own STUPID NEGLIGENCE they were accommodating...and yet this is also the one remaining thing they don't make you pay extra for. And they wonder why they are always going bankrupt.

when you come back in, it all comes back on, sort of like a giant refrigerator light.

NEGATIVE COMMENTS

While someone on the positive side noted that metering the air conditioning would make people aware of how much it costs to run a unit and that maybe they would take that awareness home with them and be more conscientious about energy use at home, the writer countered this positive with the observation that electricity doesn't really cost that much and perhaps the opposite would happen. She wondered how many people would be delighted to learn that you can run a 1000-watt air conditioner for about a quarter an hour, and conclude that electricity is one of the world's great bargains.

Someone then countered the counterer, noting that the hotels would definitely charge more than just the cost of the electricity for the metering, since less A/C use means less filter-changing and fewer repairs.

There was also a related comment describing a *Boston Globe* story about a new ultra-luxury downtown hotel opening in the middle of a recession, and how it was not likely that it would attract leisure travelers in this economic environment. Someone suggested in the comments to their story online — and referenced *Why The Heck* — that the hotel could attract leisure travelers by offering Green Rooms. I myself would happily take full responsibility for making my own bed every day (note that I said "take full responsibility for making" rather than "make" — I would no doubt leave it unmade and accept the consequences) in a really nice hotel and also not be able to take home the shampoo bottles if I could save $50 or more on a

$300 rate. Maybe other guests wouldn't but no one will know unless hotels try it, right?

PART TWO

Micro-Heckomic Ideas Requiring the Government to Actually Lift a Finger

No one said breaking free of the Methadone Economy would be easy. Yes, the proposals featured in the first part of the book wouldn't require the White House to do anything more straining than tapping out a few words on a keyboard. The proposals in Part Two, however, require a *soupcon* more effort and thought. The Executive Orders suggested here—and we are still within the realm of initiatives requiring no Congressional approval—might consume an entire paragraph or even a whole page, and would need to be accompanied by some regulations that might occasionally need updates, clarification, or correction as unforeseen situations arise.

Nonetheless, it's worth it. Part Two moves beyond the symbolic into the substantive. At this point, I say to our government employees: "Now it's time to Make History! Roll up your sleeves, open Office 2007, close Solitaire and let's get to work."

Chapter Four proposes the ShopAmerica Gift Card, which puts extra money in the hands of consumers who wish to spend it, at no cost to taxpayers by allowing them to convert tax refunds into gift cards at a substantial premium in value.

Chapter Five proposes a kinder, gentler alternative to protectionism. Products would label their domestic content percentage so that customers can consider it as a factor when making purchase decisions. This was the first idea from a non-family member to come over the virtual Why The Heck transom, and I was immediately intrigued by it. Of course, I get intrigued pretty easily. For instance, I am especially intrigued by how they get stripes into toothpaste.

The Shopamerica Gift Card:
The Grandmother of All Gift Cards

Once the White House makes those first three easy changes, each of them requiring essentially one line of text, saving money and building consumer confidence via the "Broken Stimulus" theory, it will be time to move into the more substantive *Why The Heck* proposals. Remember that *Why The Heck* is about stimulating the economy without more financial Methadone, so that even "substantive" *Why The Heck* ideas don't increase government spending.

Even so, it is probably true that the Government needs to continue to engage in deficit-spending, though hopefully at a lower rate over time, because without it, the economy ould shrink again. That's the Methadone Effect—economic overreliance on government spending.

That spending philosophy is also pure Keynesian economics. The classic example of pure Keynesian economics, from the

Great Depression, was paying men to dig ditches and then fill them in again. The Roosevelt Administration recognized the importance of getting money into people's hands, without attaching the demeaning word "handout." By contrast, putting people to work made them feel like breadwinners again. The government felt it was important to keep the work ethic in place even in the absence of productive pursuits. [10]

Why The Heck embraces the exact opposite philosophy: "Try to get more wealth and spending power into the economy without the Government adding to the deficit." It is not *in lieu of* a Keynesian Stimulus package, but rather in addition to it at first, but then gradually replacing the Methadone Economy altogether with one based on spending through tangible, environmentally sound wealth creation rather than government handouts.

While the Pitch-the-Penny Proposal, the Instant Tax Refund, and Green Rooms provide some small amount of stimulation, those proposals are about building consumer confidence that government is a source of creative innovation—the Broken Stimulus Theory. The ShopAmerica Gift Card, though also about consumer confidence, is primarily about short-term economic stimulation. In a nutshell, it gives people the option to take higher-value gift cards in lieu of cash owed them by the Government. Before I further explain it, I want to put it in the context of the Methadone Economy.

---

10. This is taught in every introductory economics course as the classic example of Keynesian economics as applied during the Great Depression. And it would be a perfect example if indeed it were true. Alas, there is not a shred of evidence that it ever happened. Apparently it was just a hypothetical used by Keynes to explain the importance of government spending for its own sake, and, as a hypothetical, was never intended to be nor was it actually implemented. Darn that Wikipedia!

The government, as noted, moved quickly to create this Methadone Economy, knowing that the alternative was worse. But consumer spending created through simply not collecting enough taxes from people to pay the bills is not sustainable. The critical question becomes: How can consumers be enticed into spending money in a sustainable way? They have to be richer.

To accomplish this, at first it was thought that a simple one-time injection would be sufficient. That theory resulted in the 2008 "stimulus checks" but most people appeared to have saved their windfall instead of spending it, and in any case no one *felt* richer because this check was a not a sustainable source of wealth creation.

But let us assume that the idea of jump-starting the economy by bribing people to spend more money was a good one in the first place. When one reviews that entire commentary around that idea, it appears that not a single economist seemed to know how to give people a bribe that they would indeed spend.

How bereft of creativity are economists on this subject? A Boston University economist proposed a government-subsidized "national 10%-off sale." This idea has a lot of problems. It would be administratively complex. It would force people to spend money to benefit. It would also create a massive peak-load problem, sort of like Black Friday every day, followed by empty malls once the sale ended, exacerbating swings in the economy.

And what would be covered? Would, for example, food be covered? People would buy the same groceries they buy now and get 10% off? No stimulus there. So maybe food is out. But what about Wal-Mart or other stores which sell food and non-food items? Also, stores have different margins.

Bloomingdale's, for instance, still makes money at 10% off and maybe shouldn't get a government subsidy for the privilege of doing so. So which stores get the subsidy? And those are just a sampling of the problems.

This was clearly an impractical proposal, not to mention one whose dollar-for-dollar reliance on government spending was pure Methadone. Yet *this proposal made* **The New York Times** *list of 2008's most interesting ideas.*[11] At the time, it probably was. The *Times* wasn't wrong – faced with a choice of solutions from economists, whose job it is to come up with them, they picked the most plausible.

And that's exactly the reason *Why The Heck* asks the following question: Who died and put economists in charge of creative solutions? Economics is primarily an observational and interpretative discipline, not a predictive or prescriptive one.[12] For instance, when Japan's economy was booming, the economics literature was chock full of articles explaining why Japan's model of public-private collaboration, consensus-driven decision-making, lifetime employment, export focus, statistical process controls, bowing from the waist, not eating pufferfish livers, and handling business cards like they are Faberge Eggs was so much more efficient than ours. But after Japan's economy stagnated, the literature overflowed with articles explaining why their model had always been doomed to failure.

My father is a psychiatrist. When I was a kid, my friends and I always assumed that he, like every other psychiatrist, possessed

---

11. "8th Annual Year in Ideas," *The New York Times Magazine* December 14, 2008
12. There are of course exceptions—some economists do predict, often very accurately. For instance, Professor Robert Shiller of Yale predicted that the dot-com bubble would burst but was largely ignored at the time. Subsequently, when he predicted that the housing bubble would burst, everyone remembered how he was proven right the first time, so he was largely ignored again.

the ability to read our minds and tell whether we were lying just by looking at us. If that had been the case, he could have made far more money by playing three-card monte in Central Park than by seeing patients. Likewise, receiving an economics degree does not automatically make you financially brilliant. Take Harvard, for example. It has one of the best economics departments and yet in early 2008 none of its brilliant economists walked across Harvard Yard to tell Harvard's money managers that there was about to be a Really Big Recession and that they should sell stocks to prevent a third of Harvard's endowment from being wiped out. To put it another way, when was the last time you met a rich economist?

There is even a joke that economists themselves tell: "The definition of 'economist' is someone who, upon learning that something works in practice, wonders whether it will work in theory."

If economists aren't going to solve our problems, someone else has got to do it. That is why www.WhyTheHeck.com pays people for ideas, and that is why it attracts ideas like the "ShopAmerica Gift Card." The ShopAmerica Gift Card solves exactly the problem that has baffled oodles of economists: getting people to spend more of their money without reliance on more government Methadone.

Here's how a ShopAmerica Gift Card would work. When they are due to receive a tax cut, a stimulus check, an income tax refund, or even a Social Security check, people would be given the *option* of either taking the check owed to them or else going to a website commissioned by the Government that shows a list of sellers in their area offering their own gift cards. Those gift cards, for many reasons, will have a face value somewhat higher than the check. Logistically, once a consumer agrees to the card, the Government cash gets credited to the seller's account and the buyer gets the card instead.

One very important point worth mentioning again because some people who've already read this have overlooked it even though it seems perfectly clear to me: Congress does NOT need to authorize additional tax cuts or stimulus checks for this proposal to work. The ShopAmerica Gift Card can be applied to *any* government check, even checks as mundane as tax refunds. So implementing this offer does not require any additional government spending or any act of Congress.

An example might be helpful. Suppose, for example, that the Government owes you $500 as a tax refund. Instead of just sending you the $500, the feds direct you to a website where a long list of sellers are offering a swap for proprietary gift cards worth more. Suppose you like Fuddruckers because they have the "Best Fuddruckin' Burgers in Town^{tm}" and they are offering $600. You receive a gift card for $600, which must be spent there, over time.

It's that simple. The government benefits as follows:

- The extra stimulus comes at no cost over and above what the Government would spend anyway;
- The "multiplier effect" on the spending would start out at a higher level because the denominator is the value of the gift card, which is greater than the size of the tax cut or other payment;
- By setting expirations (if any) the Government could time its stimulus. Varying expiration dates strategically could smooth seasonal spending patterns; and
- Money is more likely to stay in the country when people spend it. This is subtle but important: Higher-margin

sellers, such as restaurants, hotels and manufacturers, will offer higher premiums on the gift cards. Higher-margin sellers are also likely to source more of what they sell domestically. Retailers that source from low-cost countries like China tend to be lower-margin themselves and thus are likely to offer lower premiums on their gift cards, making them less attractive to consumers. So, without any protectionism, most of the money will stay in the country.

Note that there is no extra "Methadone" involved here. The ShopAmerica Gift Cards simply help the current Methadone supply reach the economy's bloodstream. Like any other medicine, less Methadone is needed if the dosage is administered more efficiently.

Any retailer, manufacturer, or service provider that passes a financial screen set up by the Government's ShopAmerica Gift Card website contractor could offer its own gift card. The seller would receive two major benefits:

- The ShopAmerica Gift Card offers very efficient, targeted marketing. Sellers could vary the gift card premium by region or zip code to direct traffic to particular stores in the chain. And unlike a subsidized "national 10% off sale" which would need a whole new bureaucracy, most sellers have experience issuing gift cards. Some retailers, like Sears, took the initiative and offered to exchange the 2008 Stimulus Check for a gift card with 10% more value, even without a formal government program; and

- Sellers can "borrow" from the Government at no cost. As with any prepaid gift card, sellers get their money before the consumer spends it. In this credit environment, having the money in advance is a huge advantage, on a large scale.

The program would be open to every financially qualified seller, regardless of size. Smaller sellers would probably group together, like mall tenants do today. For instance, an entire town's business district might offer one card that would be valid anywhere in town.

For consumers the advantage is obvious. Anyone planning to spend money will have more to spend. The title of this chapter is "The Grandmother of All Gift Cards" because think of who used to give you "free money" to spend on your birthday? Your grandmother, that's who. You'd open up her birthday card all the way and shake it to see what fell out of it. That's what this is—free money, assuming you spend it.

But it's *sustainable* free money…an efficiency which will indeed increase America's standard of living. And because wage earners are more likely to get government checks than high-income earners, this idea will disproportionately help those who need more help.

Anyone considering spending the money will see that the relative utility of spending vs. saving suddenly makes spending look very attractive. And there is nothing heavy-handed about it. Consumers who don't want to participate don't have to.

Clearly the economy periodically needs a stimulus that will actually work through the system quickly. To provide such a stimulus requiring only voluntary participation by sellers and consumers—at no cost to taxpayers—would be a boon to everyone.

COMMENTS

While every proposal in this book was available for comment on the blog, the ShopAmerica Gift Card was also "floated" in the *San Francisco Chronicle* to see if I was overlooking an obvious deal-killing objection, and in general to see what level of public acceptance could be expected. The comments in the *Chronicle* and on the blog were very positive—high acceptance, low skepticism and, most importantly, no one pointed out any obvious, invalidating deal-killer.

POSITIVE COMMENTS

The best positive comment was to point out that the government could do something similar to ShopAmerica Gift Cards for tax returns where money is owed. Filers who owe money often wait until April 15 to file, thus forcing the government to borrow money until the returns come in. Perhaps these same retailers might be convinced to offer discount coupons for those filers who owe money and file early. (Obviously, they would offer discount cards, not gift cards, because no money is coming into them to finance the cards.)

Several people pointed out that not just Sears but several other retailers also had offered to exchange the 2008 stimulus check for a gift card worth 10% more on their own. That validated the idea that retailers would participate, but 10% isn't much of a premium. However, another way to look at this is that these retailers had been offering a 10% premium under the following circumstances:

- They had to pay to advertise the offer themselves;
- They had to do the stimulus check-for-gift-card exchanges manually;

- Few retailers, and no manufacturers or other companies, were competing against them with similar offers;

- There was no central website where consumers could compare multiple offers, which would presumably make the offers more competitive in and of itself, like any other central market;

- Credit was much more available to retailers so the benefit of getting the money in advance was much lower;

- They were doing this on a small scale;

- They were doing it when stores were nowhere near as empty as they have been in 2009 and 2010.

Revise all of those circumstances to take today's economic situation into account, and one can easily imagine much higher premiums being offered by Sears and other sellers and manufacturers. Remember, not every retailer, manufacturer, or restaurant has to offer high premiums. The consumer can pick only one gift card so all a consumer needs is one attractive option. Some sellers might not even participate at all. But others will indeed offer high premiums, particularly manufacturers who operate on higher margins but with more fixed cost and likely more excess capacity. For them, moving product quickly is the primary economic imperative in hard times.

The other positive comments suggested a few variations on the gift card theme. Staggering expiration dates so that the spending could be seasonal and not duplicate the holiday shopping rush was one such idea.

Someone else pointed out that a perfect place to implement this would be to create an "Instant ShopAmerica Gift Card Tax Refund." Instead of having a check deposited to their account as described in Chapter Two, someone could ask for a ShopAmerica Gift Card instead, and get more purchasing power instantly from the tax refund than they would have gotten conventionally in a couple of weeks.

NEGATIVE COMMENTS

Several negative comments helped to refine the chapter, and aren't repeated here. The ones below describe objections which can be rebutted.

On the negative side, someone said that these cards would only move spending around, not create new spending. So someone intending to spend $500 at Target might see a Wal-Mart card offering a 10% premium (and a low-margin high-volume retailer like Wal-Mart would probably offer only 10%) and decide to go there instead. Certainly, not all the gift card spending would be new spending, not by a longshot. But an economist would say that these cards in general would "cut the cost of spending money." Anytime you cut the cost of anything, you encourage people to do more of it. This is no exception.

Yes, I know what you're thinking: "Why is he using economists to support his argument? This chapter just said that economists don't do anything." Well, that's not actually what the chapter said. It said, and I quote and you can look it up if you don't believe me: "Theirs is primarily an observational and interpretative discipline." The point about cutting the cost of spending is an observation about a fundamental law of economics. There are certain laws of economics that are strictly enforced. One is that, other things being equal, people do more of something if it's cheaper. In this case making spending cheaper than savings will encourage more spending.

And what is wrong with letting someone spend $600 at Wal-Mart instead of $500 at Target, or vice-versa?

Another negative comment cited the possibility that there could be fraud and abuse. This commenter must think that other government spending programs have no potential for fraud. That might be true on his planet, but here on Earth any time the Government spends money, some of it gets wasted or ends up in the wrong hands. To use just one example, law firms are creating practices to push stimulus spending to their own clients, rather than let the money go to the organizations that actually deserve it but that might not be able to afford high-priced lawyers to advocate for them.

In general, there is much less potential for fraud when people are spending money themselves than there is when the Government spends it, so fraud would be lower with these ShopAmerica Gift Cards than with almost anything else other than straight tax cuts and stimulus checks, neither of which stimulates the economy as much.

There is certainly the potential for a fraudulent company to set up shop, offer a high premium on a gift card, and then just collect the proceeds and skip town. However, the Government could easily contract with someone to police the gift card website. Any experienced overseer could design an algorithm that combs the gift card offerings looking for high premiums, low credit ratings, and/or limited records of conducting business, and then investigates suspicious businesses accordingly. And like most sites, there could be user ratings too.

Fraud is not limited to government programs. The most fraudulent companies in the recession were those offering cash for jewelry and gold. Ironically, nothing can be done to stop them since they do indeed pay people for these items, which is all they advertise that they will do. They just don't pay them much. The government would have much more recourse against people offering fraudulent or counterfeit ShopAmerica Gift Cards than against these sharks.

Another comment pointed out that the much of the money spent at retail establishments would leave the country. This is true to some degree, but there are three reasons why that argument does not invalidate the idea:

- Take that objection to an extreme and you wouldn't want anyone to spend any money at stores, since many consumer products are imported;

- Stay tuned. Chapter Five will propose encouraging companies to put domestic content percentage on their products, much like the EnergyStar label today, so that people who didn't want to spend money on products made abroad would have the information they need to make that choice themselves;

- Restaurants and manufacturers and service providers could also offer a gift card. They tend to have higher margins than retailers so they could offer a larger premium. Much less of that money would leave the country right away.

An economist wrote in to ask why the retailers couldn't just swap checks for gift cards themselves. He asked: "Is there really enough of a market breakdown to merit government intervention?"

Good question, so let's be clear. The Government isn't really "intervening" in the marketplace at all. The Government wouldn't necessarily *run* the website. The Government would put out a Request for Proposal to find one or more organizations willing to run such a website(s)...and possibly even pay the Government for the privilege of doing so, on the assumption that they could sell advertising. The fact is that anybody could start a website, attract retailers, and try to get consumers to send them money that then gets translated into gift cards. The Government would simply be contracting with a site(s) that it trusts to automatically turn its refund or stimulus checks into consumer gift cards. It's no different from any other free market where the Government is a participant, though as a practical matter whatever company wins the ShopAmerica Gift Card website contract will enjoy a huge advantage in efficiency over a company trying to offer a gift card service directly to consumers.

The final negative comment was a philosophical one: This card encourages people to spend more money, which is partly what got the country into that recession in the first place. The card does indeed encourage spending, and we all remember a certain president telling us that the best way to defeat terrorism was to keep shopping. Nonetheless, it is not helpful to tell someone who just got diagnosed with lung cancer that he shouldn't have started smoking in the first place. And you never see first-responders to an accident scene buckling victims' seat belts.

CHAPTER FIVE

Starting a Kinder, Gentler Trade War

During the Depression, the Government implemented some shortsighted programs and policies whose lessons are well worth studying.[13] One of the latter was the Smoot-Hawley Tariff. The idea was that in order to employ more people domestically, the country needed to keep imports out, which in turn required raising tariffs. Naturally, the countries exporting those goods to us did not cotton too well to that policy, so they raised *their* tariffs until the next thing you knew, we had a full-fledged trade war. Over a period of several years, international trade fell by two-thirds. Or maybe it was three-quarters. Whatever it was, it was a lot. Beyond that, I don't know. Perhaps one of you could look it up and get back to me on it. One way or the other, some economic historians feel the effect of this trade war exacerbated and prolonged the Depression right up to the eve of World War II.

---

13. And unlike the last one we described (ditch-digging), the one we are about to describe has, in the immortal words of the great philosopher Henry Kissinger, the additional virtue of being the truth.

These days, raising tariffs is pretty much off the table. Among other things, we can't afford to get China upset with us because if we are the spending addict, they are the enabler. However, there is an increasing chorus of whining about how we need to exact retribution for unfair trade practices, how our jobs are going overseas, etc. It's not just talk. The 2009 Stimulus Package contained a "Buy American" requirement that public-sector purchasers buy certain things like steel domestically, which will raise the cost of public projects considerably.

The bottom line is that tariffs may not be going up, but a lot of other protectionist actions and sentiment seems to quack like tariffs.

Hold that thought. Now consider two other observations.

Observation #1: Food labels have listed ingredients for decades. Some people ignore them and others read them. No one questions the value of the information to the consumer, though. No one says "This information shouldn't be there. People should have to guess."

Observation #2: Appliances have been sporting EnergyStar labels for decades. Once again, some people ignore them while others don't, but no one questions their value and thinks they shouldn't be there.

Now take that initial thought off "hold" and consider it in light of those two observations. Some people are very interested to know which products are made in the United States. So why can't all consumer products list their domestic content percentage? And why can't stores list their overall domestic content percentage? Some people will ignore the information but others who prefer to "buy American" will use that information as much as possible.

Is this protectionism? It does not involve tariffs, quotas, hidden subsidies, government rules about public-sector purchasing, or even incentives. It is purely informative. If other countries "retaliate" by listing their own domestic content percentages, so be it. It is protectionism only at the consumer's option. Only people who want to alter their purchase decisions to buy American-made products, will do so. Domestic manufacture is a product feature, the same as any other product feature, and it is certainly not the silliest thing people pay extra for. That award would probably go to $15 service contracts on $39 phones.

And then there is the whole rental car experience. How many times have you bought a full tank of gasoline and used only half of it? Initialed "yes" on all eleveteen of those insurance coverages even though you're 99% sure that your own insurance company would cover you anyway? ("This one covers your deductible if your special sauce spills onto your two all-beef patties before it gets absorbed by your sesame-seed bun.")

In concept the idea of domestic content labeling is quite simple. In lieu of tariffs or explicit governmental interferences in the marketplace, just let people decide whether they would like to support American manufacturing, or, in other words, whether they want to slap a tariff on themselves. The first description sounds very supportive of consumers who would like to buy American, while the second makes them sound like idiots. That's exactly the point. If you're in the latter camp and believe it is foolish to pay more for American goods, you are free to ignore the information.

Next comes the question of whether to make compliance with domestic content labeling mandatory or voluntary. The philosophy of *Why The Heck* would generally oppose making anything mandatory. Instead, someone could set up some standards for reporting and start a campaign to encourage manufacturers and retailers to comply. No doubt the ones that source primarily domestically will jump at the opportunity. This might force the others to go along on the theory that people will assume the worst if they don't.

Mandatory domestic content labeling would have the advantage of 100% compliance, but initially some businesses with low domestic content percentages would lobby against such a provision, so it might take a while to implement. Another option is to make labeling mandatory for a year and voluntary after that. It is unlikely that many companies would drop out of the labeling program if consumers are responding to it.

Or the program could institute a preference for labeling for consumer products purchased by the Government, or any state government accepting Stimulus funds. Then you are doing it on the "buy-side," via an administrative order, so that compliance would be completely voluntary even among companies selling to the Government. Standards would be published, but no laws would be needed.

Of course it is possible that other countries would "retaliate" and create a virtual trade war by listing their own domestic content to encourage their own citizens to buy their own products. Here at *Why The Heck* Corporate World Headquarters, we say, let 'em. Domestic content is a product feature, and knowing whether a product has a lot of it is a useful piece of information no matter what country you live in.

There would likely be two unintended consequences. First, support for the next auto industry bailout will evaporate. The average person has no idea that the difference in domestic content between many foreign autos and many Big Three autos is not even close to 100%. In automobiles, there are few totally "domestic cars" and only a few more totally "foreign cars." The difference in domestic content is probably more like a third, on average. That means if Chrysler and GM ever do stop making cars altogether, the majority of their foregone auto production will remain in the U.S. even if the actual brand names of their cars change.

Second, there are some companies, notably also in auto manufacturing, that have plants both in the United States and elsewhere. They move production around depending on currency fluctuations, capacity utilization, etc. Domestic content labeling makes it more likely that, at the margin, they will source more production intended for sale in the United States right here in the United States.

One final observation: There *already is* domestic content labeling for automobiles. I waited until the end to mention it because I wanted to see how many people already knew about it. Don't feel bad. It's not like the EPA Mileage Estimates, which scream at you. Domestic content is listed on the part of the sticker where they list the wheelbase, curbside weight, and cubic liters of engine displacement, along with the list of standard features including the ultra-deluxe bi-level remote-controlled calfskin-lined power-assisted cupholders, and all those other things no one reads, apparently including you. It's just a question of making domestic content labeling more prominent and extending the concept to all consumer goods.

To get an idea of what this labeling might look like in practice, read one of these stickers, and then imagine it bigger and more prominent, like an EnergyStar sticker. Since they are only found on the car window, you'll have to go to a local dealer, maybe one who still sells Chryslers, to do this. The dealer won't mind. He'll be happy to have the company.

Haha, good one, Al. In all seriousness, this is no joking matter. Traffic at many Chrysler dealers is so slow that a lot of dealers are diversifying into other, higher-volume, lines of business to increase it, such as repairing Maytags.

COMMENTS

Comments on this posting were uniformly positive with one note of caution.

POSITIVE COMMENTS

There was a bit of an intellectual and/or philosophical and/or semantic and/or boring debate about whether this proposal constitutes protectionism. Consensus seems to be that it doesn't because it does not erect barriers.

Two people said that they would alter their spending patterns if they were privy to this information. Make that three, including me. Once I narrowed my choice of a product down to a few finalists, I would opt to "Buy American." Keeping my fellow countrymen employed is important to me because I like to think of myself as an empathic person, so wherever possible, I consider the needs of others. This way, those less fortunate than I could afford food, shelter, and the occasional splurge on a pop economics book. This would be especially true in a product category where I was fairly confident the American-made offering wasn't unreliable junk likely to break down on me as soon as the warranty expires, such as coal.[14]

Someone else suggested that the concept of domestic content labeling be applied to the ShopAmerica Gift Card, meaning that some kind of listing preference on the ShopAmerica Gift Card website would be given to manufacturers that agree to list their domestic content percentage on their products. The domestic content percentage for that manufacturer as a whole could appear right on the website.

Yet another comment suggested applying this concept to food production not just by country but by state, in order to gently encourage more consumption of local foods.

The only negative comment was a caution that I would share: Make sure that only one number—the actual domestic content—is listed. Don't start listing percent-of-content-by-country. Some countries might take it personally and think we were drawing undue attention to subtle nuances like the fact that half of their products are poisonous.[15]

---

14. Okay, that was a cheap shot and I didn't really mean it but I couldn't resist. Fact is, my car was assembled in America and it is quite good, as are many other American-made items. We might import a lot of simple things, but we still manufacture products with a great deal of complexity in them, like aircraft, better than anyone. We also export entertainment, an industry with quite a large domestic market because we Americans excel at mindlessly amusing ourselves while our subordinates complete projects that we take credit for.

15. No names but it might rhyme with China

Part Three

Eliminating middlemen, those leeches who suck the lifeblood out of an economy at taxpayer expense to further their own evil plans, such as me

Here is a Dramatic Confession. I am a middleman.

If I were on *To Tell The Truth* I wouldn't be one of the contestants who are lying to Orson Bean and Kitty Carlisle[16] in the hopes of winning Lovely Parting Gifts. Instead, I would be the Real Middleman who would Please Stand Up. If you don't believe me (and why would I lie about something as shameful as this?), you can visit my Day Job website at www.dismgmt.com. In my case I'd like to think my middleman job is justified because no one else in my industry appears capable of counting. For instance, one supplier in my field claimed on its website

---

16. Assuming that show even still exists on some cable channel somewhere, I don't actually think Orson Bean and Kitty Carlisle have appeared as panelists since the Pleistocene Epoch, but you gotta love those names.

that their productivity-enhancement solution could "reduce absenteeism by 300%." Maybe that means that workers are so grateful to work an extra day that they put in three days of free overtime.

And they aren't the only innumerates. Sort of like the Matisse masterpiece that hung in the Museum of Modern Art for a month before someone noticed it was upside-down, the 300% claim on this company's site was posted for *two whole years* before someone—*moi*—noticed that the math was impossible. That goes to show that my competitors can't count either. In fact, I attribute the success of my day job not to any particular competence on my part, but rather to the good fortune of being able to compete against some of the most clueless executives this side of Detroit.

One more anecdote about the complete innumeracy in my field and then I promise I'll shut up about it. A presenter for a medical management vendor was showing case study after case study demonstrating how much money its programs saved for specific hospitalized patients they managed. I kept asking the presenter to provide the *average* savings per patient, and she kept saying she would get to that. Finally, after about the eighth case study, I couldn't take it anymore, and I asked her if she could pleeeeease tell me what the average is.

"There is no average," she replied. "It varies."

So I like to think that as long as I can do fifth grade math in an industry that apparently can't, I am the exception to the general rule that in an Internet society, middlemen—especially those whose fees are either hidden and/or paid directly or indirectly by the taxpayer—have outlived their usefulness. The first two chapters in this section go after middlemen whose fees fit both categories.

Chapter Six goes after the residential real estate brokerage industry, which is doing more than any other group of smiley-faced part-time workers to exacerbate the housing crisis and increase the size of various bailouts. This chapter tackles the reason it requires not one but two of these brokers to tell you to "notice this hardwood floor," as though you could possibly not be noticing it because you're standing on it in lieu of falling into the basement. This chapter has attracted a TON of opposition from real estate brokers mostly because there are just too many real estate brokers with enough extra time on their hands to go around commenting on blogs, which indeed is part of the problem.

Chapter Seven observes that Medicare — meaning you and me as taxpayers — shouldn't be spending money on brokers whose job it is to get seniors to enroll in their HMOs. I don't much mind paying taxes to support health care for seniors, but I really do mind paying taxes to support healthcare brokers for seniors. Once again, the size and decibel level of the opposition has convinced me that I am onto something.

Why the Heck Aren't we Already Doing this Stuff

CHAPTER SIX

Time to Blow Up the Real Estate Brokerage Cartel
(And Find A Way To Notice Those Hardwood Floors All By Ourselves)

The introduction listed four goals of *Why The Heck* proposals, and reconfiguring the real estate brokerage industry fits all of them.

News flash: Residential real estate commissions do matter a lot to the economy, far more than anyone has suggested, to the extent anyone has suggested the connection at all. As an analogy, look at stock commissions to see how transactions costs and illiquidity can disrupt an entire economy. Thanks to deregulation of commissions, the stock market has become "liquid" enough that people can get out of stocks much more easily and inexpensively today than in the past. The immediate trigger for the 1929 crash was largely "margin calls," requirements that people who had bought stock primarily with borrowed money put up more money as stock prices fell.

Yet despite comparable stock price declines during the 2008-2009 bear market, margin calls were rare. People can no longer buy stock with little money down. Equally important, they can easily sell margined holdings before their equity goes to zero and they must pay the brokerage firm to cover their margin borrowings, because the stock market is so liquid and transactions costs are so low. The happy result: Even during the recent calamitous decline in the stock market in which investment firms found many other creative ways to fall on their swords, no stock brokerage firm went bankrupt because people faced margin calls that they couldn't pay.

Can you just cut to the chase? Some of us don't have time to read your too-clever-by-half book. It takes forever even to read just the *title*.

Fine. Be that way.

To boost housing sales, the government has tried every way to influence the market except one: facilitating it. The way to facilitate the market is to make buying and selling houses much less expensive.

Sales increase if transaction costs decline. Lower transaction costs make it more attractive for short-term investors to buy, rehab, and sell distressed real estate. Also, lower transaction costs increase a seller's equity while reducing a buyer's purchase price. Lowering the buyer's effective purchase price will entice more long-term buyers into the market. That was what the Government did when they subsidized homebuyers during 2009-2010. The Government can accomplish the same thing, at no expense, simply by being smarter.

For many reasons (and, yes, you need to read the chapter to learn them), real estate transaction costs are artificially high, exacerbating the ongoing residential real estate malaise. Both states and the federal government could easily create pressure to reduce them.

First, states could reduce commissions by allowing prospects to walk into real estate offices and say "I am acting as my own broker," and be entitled to the buy-side broker's half of the total brokerage fee at closing. The realtors would have to list two prices: one for people who come with their own brokers and one for people who don't. Once prospects see what a broker really costs them, they will house-shop on their own, and/or retain a lower-priced advisor.

Next, a large portion of all first mortgages are federally financed nowadays. The federal lending agencies could simply reduce the down payment for buyers whose purchase prices don't include a full real estate commission. Transaction costs do not add to a house's value. Therefore a lender could get greater default protection by requiring a lower down payment on a lower-priced house purchased directly from the owner than a higher down payment would provide on the same house with a commission added onto the price.

Once buyers – who are currently completely insulated from physical transaction costs and hence unlikely to pressure them to decline – realize that the size of their down payment check depends on the size of the seller's brokerage commissions, they will flock to the web seeking for-sale-by-owner listings. Once owners see buyers home-shopping *en masse* online, most will, themselves, migrate to the web. Replacing 5-7% commissions with internet-level transaction costs (supplemented by advisory fees for the seller to help in marketing/showing) means higher proceeds for sellers but lower prices for buyers…a sure way to turbocharge sales AND reduce foreclosures.

Web entrepreneurs would accommodate this shift, charging buyers and sellers internet-level fees to be matched up. The archaic two-broker model will be supplemented and ultimately supplanted by much less expensive advisory, marketing and appraisal services to support homebuyers and sellers.

Of course, people who want full service brokerage could continue to get it, whether as sellers or buyers. The difference is that full service becomes an option, not a *de facto* obligation. Meanwhile, the rest of us will share in the benefits of low transaction costs – smaller down payments, additional choices, faster and easier sales, more mobility, fewer foreclosures, and ultimately extra money in everyone's pockets.

That's not the case with real estate. Real estate rules today are eerily parallel to margin rules in 1929. Almost anyone can still buy a house for 20% down. (By contrast, buying stock required 10% down in 1929.) The day you move into a house, having paid a 6% brokerage commission and two "points" on the mortgage, plus closing costs, moving costs and taxes, you are already about 9% in the hole. So if you lose your job a day later, you can't just move out. Your 20% down payment, less that 9%, now nets out to an 11% equity stake.

Add to that 9% haircut the risk of a decline in housing values and a few delinquent mortgage payments, and it's easy to see how people's equity can be wiped out far, far faster than home values decline, and why a lot of nervous buyers are sitting on the sidelines to avoid the possibility of that happening to them. The cliché people often use to describe buying a house, "I'm going to take the plunge," turns about to be almost literally true. Your equity plunges the day you move in.

If total transactions costs were much lower, buyers would not endure remotely as great a first-day plunge. Many more people "stuck" in their homes would be able to cut their losses and get out. Many more potential buyers now on the sidelines would come into the market. In short, the market would become more liquid, more efficient, more forgiving. Houses could pass much more cheaply and quickly from those who can't afford them, including developers with big unsold houses and big unpaid bank loans, to those who can. Fewer bankers would be "stuck" with unsold houses.

I don't expect you to have sympathy for bankers, but foreclosures are a significant drain on the economy (not to mention on the people who are foreclosed upon). So, reducing transactions costs to increase liquidity and allow orderly sales would be a huge benefit to the banking system, as well as to every homeowner, whose equity in their own house increases.

Did real estate brokers cause the housing crisis? Let's put it this way. If Hollywood makes a movie about the housing crisis, the brokerage industry might receive a "Best Supporting" nomination. As one of the comments points out, the two categories of "good broker" and "successful broker" are only tangentially related. A good broker would never recommend that clients buy houses they couldn't afford, while, by definition, many successful brokers were doing exactly that.

Even if one assumes the opposite—that a broker would rather go hungry than sell a house to someone who couldn't pay for it—the "answer" going forward doesn't change, because oftentimes solving a problem requires an entirely different protocol from preventing it. For instance, you wouldn't try to prevent someone from breaking a leg by putting a cast on it.

Will much lower brokerage fees eliminate the need for most foreclosures? Of course not. A problem that triggers the economic meltdown of an entire planet has multiple causes and no single solution. A large reduction in transactions costs, creating a similar increase in liquidity, would be a substantial part of that solution even though it would not solve the problem altogether.

While only a partial solution, the best thing about this part of the solution, this being *Why The Heck*, is that it is free. Other "solutions" for housing are all about moving mortgages around, subsidizing down payments, or having the Government take them over. *Why The Heck* takes no position on all that stuff— solutions requiring subsidies and bailouts are outside the realm of *Why The Heck*. *Why The Heck* contributes creative solutions. It does not advocate for or against taxpayer-financed relief of homeowners or bankers. We make enough enemies as it is.

Nor is it enough to say that brokerage fees are too high. In the broadest sense, a lot of prices seem "too high" and should be much lower. Take tomatoes. When I was a kid, they cost about

49 cents a pound. (Not that I ever ate one, my diet consisting mostly of Captain Crunch.) Now they are more like $2.79. That price reflects the high costs of picking and processing, a very labor-intensive process now performed by relatively well-compensated farmworkers. So you can say, "The price of tomatoes should be lower," but those prices are high because decades ago many well-intentioned people, including me, pressured growers to pay farmworkers a living wage. There is no monopoly on tomatoes because there are tens of thousands of growers and a functioning market.

You cannot just say that a price "should be" much lower. The price could be high because, as with tomatoes, costs are high. You'd need to show actual reasons that the industry is a price-fixing cartel rather than a regular marketplace. And because, like tomato growers, there are tens of thousands of real estate brokers, it's not immediately evident that this industry could constitute a monopoly or cartel. Instead it will require a lot of ink to prove it.

A lot of ink indeed, so let me first apologize to those readers who expect that every *Why The Heck* idea can be described in 1500 words or less. The real estate brokerage industry has gone to a great deal of trouble to maintain its commission structure, avoid government scrutiny and resist competitive pressures despite a tidal wave of democratized information, transparency, and the Internet, a tidal wave so powerful that it would have drowned King Canute himself.[17]   It will take more than 1500 words to explain how to undo decades of price-fixing. To make it more digestible I will use italicized subheadings in this chapter from

---

17. King Canute, as everyone knows, thought that his royal powers allowed him to hold back the tides. You might wonder, "What was this guy thinking?" and you'd be right. It turns out that he would routinely go down to the shore to prove the opposite—that he couldn't do this. Darn that Wikipedia.

popular songs. If you get bored with the exposition you can see if you can recognize the lyrics.

In '69 I was 21

In 1969, selling $100,000 worth of stock cost $2000, and selling a $100,000 house cost $6000. Suppose the value of each asset has since quadrupled. Today, selling a $400,000 house costs $24,000 while selling $400,000 worth of stock costs about $400, and even then you'd pay $400 only if you didn't shop around.

In 1969, you used one broker for either type of transaction. Today, most stock transactions involve no live human brokers, but most residential real estate transactions involve two.

Stock commissions have plummeted. Commissions for everything else have fallen too. Dealer markups on cars would be another example. Widespread availability and transparency of auto dealer economics have driven markups on all but the most popular models into the single digits. And after former New York Governor Eliot Spitzer, who will now be remembered exclusively as a punchline, shined a bright light on commission practices in the insurance business, they atomized. Here in Massachusetts, deregulated auto insurance is reducing hidden broker commissions, producing significant savings for consumers.

So everywhere in this great but troubled land of ours, markets are becoming more efficient, buyers are enjoying better deals, and middlemen are getting their resumes out, victims of the inexorable trend towards transparency and economic democracy generated by the Internet.

Everywhere except residential real estate. And yet these are the commissions that matter the most in this economy. Why?

Because we taxpayers are indirectly paying these commissions in the form of bailouts and foreclosures. Absent these bailouts and foreclosures, the efficient functioning of this market matters only to homeowners, home buyers, and potential home buyers. Hey, wait, isn't that almost all of us too?

That something's wrong here, there can be no denyin'

There are at least nine reasons why the real estate brokerage industry simply does not pass the sniff test as a competitive marketplace. Finding nine reasons is huge — history is full of examples where monopolies were broken up based on one or two violations of antitrust law. Really, only one or two of the reasons below would suffice, but here is the full list:

1. All the brokerage firms in an area post the same fees (except the rare discount broker). Imagine the outcry if auto dealers or supermarkets did that.

2. While, as mentioned, a few firms charge less than the "standard" commission, no firm charges a higher rate and justifies it by claiming that their level of service is so high, their ability to get the best price so unparalleled, that a customer should be willing to pay more to use their services. They know that the "standard" is as high as they can charge.

3. The fees vary by state and occasionally by town, but rarely do they vary by brokerage office within the town. One of the commenters on *Why The Heck* lambasted me for saying that commissions were 5% in Massachusetts. "In Medford they are 4%," she said. So if I buy a house in Medford, I pay a 20% lower commission than in the other 350 cities and towns? What makes Medford so

special? What if this were true for auto sales? What if a car salesman asked you what town you lived in and said, "Good for you. That means you can buy a car for 20% less"? Obviously that would never happen in a functioning market, but it does in real estate.

4. In law and medicine, fee-splitting is illegal. In residential real estate brokerage, it is the rule.

5. As mentioned earlier, most transactions involve two brokers, one "representing" the buyer and the other representing the seller. It's not just the sheer number of brokers which raises eyebrows. It is the way they are paid. Both make more money, the higher the sales price, and neither gets paid until the deal is closed. Can anyone spell "conflict of interest"? The mantra of *Why The Heck*: "It does not take two brokers to tell you to look at those hardwood floors. It shouldn't even take one."

6. Walk into a brokerage storefront alone and buy a house from that broker, and the broker will gross maybe 6% (the number varies a bit depending on the state). Walk into a brokerage storefront accompanied by another broker, and the commission is still 6%...but now it is split and the broker with the storefront grosses only half that amount, despite doing the same work, or maybe doing more work because he or she has to "negotiate" against "your" broker. Name any other business where the price is halved depending on who happens to walk into the storefront.

7. And, speaking of storefronts, why are there always storefronts, usually in the nicest part of town, for real estate brokers? Years ago, there were also storefronts for stockbrokers too. You know why? It's not that a

brokerage office is a relevant consideration in buying stock. They opened offices because they were making so much more money than they should have been making that it made economic sense to build a whole bricks-and-mortar storefront to attract "walk-ins" who might buy stock. Now, of course, a stock brokerage storefront is a rarity because commissions are market-driven and the industry is highly competitive.

8. Licensing requirements are onerous. Showing houses is not what a colleague of mine once referred to as "rocket scientry." Yet you need to become an apprentice, and then an agent, and then finally you could reach the level of being able to open an office. Call me a conspiracy theorist (and admittedly I do think there was a second gunman on the grassy knoll), but it seems like the point of the licensure is to keep people out who want to just sell their own homes or homes for friends.

9. The industry cannot explain why other developed countries have 2-4% commissions despite generally lower housing prices.

No such thing as the real world

Next, let us explore why residential real estate brokers can somehow continue to operate in a parallel universe where laws of economics don't get enforced. Once again, stock brokerage can be used as an analogy.

In both the securities and real estate brokerage industries, information was once available only to the broker, who would share this information with his or her clients. In real estate that would include comparables and information on the neighborhoods and so on, but mostly the "information" was that they had their own listings. They could charge a monopoly fee because you could see a house only by going to the listing broker — that broker had a "monopoly" on that house. Today, all information about the neighborhood and almost all listings are online, just like a great deal of information is online for securities. If anything, the democratization of information has taken place much more thoroughly in real estate than it has in securities. There is no concept of "trading on inside information" or the fear that some well-connected home buyers systematically know more than you do.

Continuing the parallel, both securities brokerage and real estate are highly competitive industries where many companies compete for your business. Add that structure to the democratization of information, and you'd expect that commissions in both fields would be under equally tremendous pressure. Like securities brokerage, real estate should be as efficient a market as exists anywhere, with transactions fees just high enough to cover true economic costs and a small profit.

And yet real estate brokerage fees have barely budged even as securities brokerage fees have almost disappeared. Why is the commission so much higher and "stickier" for real estate brokerage? We know it doesn't reflect high underlying costs because there are so many residential real estate brokerage offices and so many brokers and the average real estate broker sells only a few houses a year, and the cost of advertising a house's availability has come down dramatically since the Multiple Listing Service went online.

Clearly, this business attracts far more participants than it would if its pricing fell to a level appropriate to its costs. How do we know? Well, as mentioned above, how many other types of transactions require two brokers?

Yet, unlike securities brokerage, real estate brokers still informally appear to enforce a strict standard commission among their brethren. Thousands of brokers coincidentally charge exactly the same commission. This practice would almost certainly be found to be illegal because market forces without interference would make real estate fees behave over time much more like stock brokerage fees. Even so, no state attorney general has brought a case against the industry. Taking on the industry would be a politically questionable undertaking given the lack of public outrage against it. There is no evil face of real estate brokerage like Enron's Ken Lay or Tyco's Dennis Kozlowski or Wall Street's Bernie Madoff or baseball's Roger Clemens. Rather, it is mostly just a bunch of part-time empty-nesters. And statistically speaking since there are so many of them, one or two of them probably live on your block. Also statistically speaking they are likely to be very nice neighbors since they want your listing someday.

Niceness is not a defense in an antitrust suit, but when combined with its low profile and its hidden fees, niceness helps explain why the industry has largely escaped the kind of public outrage given to Enron, Tyco, and just about everyone who works on Wall Street except maybe the guy who sweeps it. The closest the industry came to being held accountable was that in 2007 the Justice Department entered a consent decree making the National Association of Realtors agree to adhere to their own standards, which would be like making a drunk driver agree to adhere to the standards of driving expected from drunk drivers.

Bottom line: There has been no serious government scrutiny of industry prices practices. Yet even in the absence of government scrutiny, one would think the sheer number of realtors and increased transparency would prevent price collusion, so how can this industry still enforce its pricing discipline?

First, *the industry artificially controls the channels of distribution.* Even though anyone can see the listings, the real estate brokerage industry owns the Multiple Listing Service (MLS). While homeowners can get their own house listed on the MLS for a sizable fee without using a broker, that option is poorly understood and what buyer's broker would show it to their clients if there was no commission potential for the sale?

Second, *they can punish those who stray.* Stock brokerage firms compete with each other. A stockbroker would never have to deal with another stockbroker to move stock. There are central markets for transacting stock, like the New York Stock Exchange or NASDAQ. The nature of the real estate brokerage business, which has changed little since the (previously non-computerized) MLS was introduced, involves both a buy-side and a sell-side representative, paid out of the same 6% either literally or functionally. (Aside from some possible liability issues, it makes little difference to the broker's economic incentives whether the broker is formally representing the buyer or not. Either way, they don't get paid until the deal gets done.) Because the same realtors deal with one another all the time, it is easy for some realtors who represent buyers to "punish" brokers who stray from the standard commission by simply not showing their houses, or showing them unenthusiastically. The brokers deal with each customer once, but with other brokers many times. Whose lasting impression do you think is more important to them?

Consider an analogous situation in health care. Two decades ago many health insurance companies instituted "second surgical opinion" programs in an attempt to cut down on unnecessary surgeries. The predictable result: Any surgeon who denied a colleague's surgery request often found that colleague subsequently denying his. Those surgeons soon realized that the patient was not a repeat customer. The other surgeon, however, was a *de facto* repeat customer. Denials disappeared, and those programs were largely dropped shortly after they were instituted.

Ironically the only thing that keeps this collusion from happening in real estate even more often than it does is that because the commissions are so high, there are often far too many brokers in an area for them all to know one another well enough to collude.

Because no industry voluntarily changes structure from closed monopoly to open competition, one can surmise that when the realtors decided to create MLS, they anticipated that the widespread dissemination of information could spell the end of the previous commission structure, which was economically if not morally justified back when realtors had their own secret cache of listings. And when I say "anticipated it," I mean, "put mechanisms in place to prevent it." I don't know the safeguards they put in place, but no other business has gone from closed to open and maintained its gross margin like real estate has.

Third, *lack of transparency in the fees*. The buyer generally does not know that even though the seller pays the commission, the burden of the commission "falls" partly on the buyer. This is because pricing is not transparent — both the seller's and buyer's commissions are hidden in the price of the house.   Eventually, the buyer sees the fee (at the closing) but doesn't appreciate that he is paying some or all of it via a higher price.

Fourth, *fuzzy math discourages discounting*. Discounting commissions is financially very unattractive, because each commission must be split between the buying and selling brokers. In any other industry, cutting your price by 20% reduces your revenues by 20%. Commission splitting in real estate means that cutting your price by 20% reduces your share of the brokerage fee, which is one-half the total brokerage fee, by 40%.

If two men say they're Jesus, one of them must be wrong

So what do the real estate brokers say in response? One posting to *Why The Heck*, called with no apparent sense of irony "Against The Democratization of Real Estate Brokerage," cites studies, no doubt sponsored by the brokerage industry though a formal citation was not provided, saying that 92% of sellers feel that they got a better price for their houses by going through a broker.

That's great for the seller, if indeed it's true. Unfortunately, a very detailed analysis in *Freakonomics* proves the opposite. Professor Steve Levitt's elegant deconstruction of the brokerage cartel based on both arithmetic and data finds that, much like the rest of the human race, sell-side realtors act in their own best interest, which is usually not the client's.

However, let us assume the commenter is right and the seller does get a better price. What about the buyer? The buyer and the seller can't both be getting better prices if commissions are a fixed percentage. Yet, half of the commission burden falls on each so the buyer is getting no value for his half.

In addition to that posting, note the comments at the end of this chapter. Even the ones contributed by brokers paradoxically made the case against brokers. One sample: "Realtors know which houses are overpriced and which aren't." Um, could that be because they are the ones overpricing them?

Break down. Go ahead and give it to me.

Clearly, this market is in breakdown when even its defenders can't defend it. In normal times this would be an issue only between sellers/buyers and brokers. But these are not normal times, because the Government, meaning taxpayers, is party to some of these home-selling transactions, when the house is "underwater." Substantially reducing the transactions cost could have very positive implications for the entire federal bailout if it could keep many houses out of foreclosure altogether.

The most important thing to note, which many of the myriad of comments from brokers failed to note, is that no one is proposing to outlaw business as usual. People would still be able to use brokers and pay a full commission, just like they do now. The difference is that full-service (meaning full-commission) representation should become an option, with a clearly transparent and obvious cost against which the realtors must offer a benefit, rather than a *de facto* requirement. To put it in sound-bite terms, using a broker should be a privilege, not an obligation.

Likewise, in securities brokerage, using a broker was a *de facto* requirement decades ago. Now it is an option. Buying stocks is much riskier than buying a house (assuming one has counsel, building inspector etc.). Yet people have managed to buy stocks on their own with the same results as professionals. Perhaps that sounds like when Mrs. Robinson accuses Benjamin of not finding her attractive, and Benjamin replies: "Now, that's not true, Mrs. Robinson. I find you to be among the most attractive of all of my parents' friends." Hey, it's not like real estate brokers have been making their clients rich lately either.

How could the federal government accomplish this solution, free? It would be ridiculously easy.

Simply change the down payment rules on mortgages involving the federal government in some way, a category which accounts for a sizable minority of all mortgages. Down payment requirements vary on these loans, but let us assume a 10% down payment is required. Why not require instead a down payment equal to 5% + the real estate commission included in the price of the property? So that, for instance, if the total commission were 6%, the down payment would be 11%. But if there were no commission, the down payment requirement would be 5%.

That simple step would create a lot of price-awareness on the part of buyers AND sellers because each point of commission paid in the price of the property would have to be explicitly paid to the lender in the down payment. The seller might be more willing to sell via the internet or negotiate a rate. The buyer would also negotiate commission or do without a buy-side agent, or pay by the hour. Ultimately, a buyer could buy the exact same house, the seller could receive the exact same net proceeds...and the buyer could have a lot of money left over. Alternatively, the seller might realize higher net proceeds while the buyer gets a lower price. Either way, the principals – you and me – are at least as well off if not somewhat better off than including a full commission in the price.

Lenders would benefit too because (to use the example above) a 10% down payment on a house with a 6% commission does not protect their financial stake in the property with a true financial "cushion" as much as a 5% down payment with no commission.

Some private-sector lenders would follow suit, and soon the entire commission structure would become much more negotiable for everyone.

The only – and I mean only – argument against this is that sellers will be able to make more money if buyers aren't represented by a broker, because the buyers will not be sophisticated enough to get the best price. This argument has more holes than Bonnie and Clyde:

- It's not like buyers weren't getting ripped off even when they were represented, during the recent real estate bubble;

- Because buyers' brokers only make commissions when the house is sold, many of them won't be racing to prevent buyers from making offers if the price is too high;

- Unlike everything else people purchase, there is also an independent appraisal. If the price is too high they won't qualify for a loan;

- It's not so bad if sellers make more money than otherwise – many of them are underwater;

- A lot of the time, the seller will realize more proceeds *and* the buyer will get a better deal, with a lower commission.

In addition, there is even a very easy way to facilitate brokerage fee reductions. The only catch is, it would have to be done on the state level, and the state real estate regulatory agencies tend to staffed by real estate executives. Some state legislatures seem to feel that having enough credentials and experience qualifies a fox for a job overseeing security in a henhouse. But here's what they could do if they wanted to actually help people other than their friends.

States could create an option that says people can act as their own brokers when buying houses. You could go into a brokerage firm and say "I am representing myself," and be entitled to half the commission. Maybe the objection from the real estate lobby would be that people don't know enough to represent themselves (like it takes a lot of skill to look at hardwood floors). That could be solved by an online course and test in real estate literacy to become a self-broker. And, remember, in the absence of a broker, people aren't more likely to seriously overpay for a house because they still need an appraisal to get a mortgage. Also, history would say that people *did* seriously overpay during the bubble, even with their own brokers.

To give this option some teeth, realtors could be required to list two prices, one for customers representing themselves and one for customers who bring brokers.

The law could also require that as a condition of license, any real estate broker or brokerage firm must offer a rate per hour, per listing, or per showing for prospective buyers who prefer to retain an advisor as a consultant rather than a broker. Any buyer or seller who uses a real estate consultant would get access to the MLS through that consultant. The consulting fee wouldn't have to be set by law. With thousands of agents in every state, who would no longer have to go through a brokerage office, competition would be intense. There would be no benefit to working in a real estate brokerage firm. Individual agents could offer this option on their own over the Internet. The licensing requirement of multiple fee structures could probably be rescinded in a year or two. By that time the market will have found its new equilibrium with multiple payment options and no longer need such close oversight.

Today, virtually no realtor offers multiple payment structures. I know because I tried to find one. I offered to pay $100/hour up front, along with several other arrangements that should have been economically appealing to people trying to show my house. Yet despite my eagerness to utilize their services, I got the impression that no one was interested in performing even the minimal amount of services needed to earn my money. It reminded me of the time I tried to take a train in a Red State.

A number of brokers lied and said that it is illegal for them to vary the pricing structure by rebating their half of the commission in favor of a consulting fee. Because it is perfectly legal to set one's own price, in a functioning market you would think that at least one of thousands of agents would sometimes prefer getting a smaller amount of money immediately, with certainty, vs. trying to sell a house for a brokerage fee. Today, using this approach would require an awkward accounting/invoicing and rebate at closing of part of the commission to the party (buyer or seller) represented by a consultant, who himself/herself has been getting paid by the hour or the showing. It's awkward even to write that sentence, let alone actually do it.

Of course the real estate brokerage industry will oppose all of this nonetheless, even though it's tough to argue against a proposal to simply expose commissions to light and let markets do their job. Plus, an equally powerful coalition of builders, bankers, mortgage brokers and consumer advocates will be lined up against them. The brokers don't stand a chance: In the immortal words of the great philosopher Victor Hugo, or

Voltaire or Descartes or Candide or one of those guys, there is nothing so powerful as an idea whose time has come.

COMMENTS:

This posting attracted more positive comments and more negative comments than any other. All of the negative comments were from brokers, of course. A surprising number of the positive ones were from former brokers, talking about how "these women (and yes, mostly women) are sharks, representing no one but themselves."

However, far more interesting than the justifiable outrage in the positive comments was the curious consistency with which the negative comments unintentionally made a better case that this industry is a cartel than I did. So, in this case, the "negative" comments serve both as negative and positive comments. There is no reason to list the positive ones because in this case they just agreed with the posting but with one exception didn't add much insight. The exception was a comment pointing out that part of the reason that this industry can continue to operate like a cartel is that in most states the real estate oversight body is comprised of real estate brokers, considered by some state legislatures to be the people with the most insight into the field.

The individual who posted this information about industry oversight, Joel Stern, would be happy to field reader follow-up questions at SternJ2@state.gov. His view after reading my posting is that I am way off...because it is much worse than I am describing. He is willing to share voluminous information dealing with conflicts of interest, toothless administrative remedies, and real-life examples of the excesses I describe.

An actual realtor also commented positively. His brokerage firm, SmithAdams, uses a business model like what this chapter proposes, and the realtor thinks his firm may be the only one doing so. One is an ideal number in situations like this—high enough for "proof of concept" but low enough to put only the tiniest dent in the *Why The Heck* claim of originality.

NEGATIVE COMMENTS

Members of the real estate brokerage industry contributed a majority of the negative comments. The exact percentage of negative comments attributable to brokers might be...let me get out my calculator here...hmm...works out to roughly maybe around approximately 100.000%.

They do make some excellent points about why someone might *want* to use a broker. It saves the seller time and effort, and the broker is likely to be good at writing listings.[18] (There were no comments explaining why a buyer would want to use a broker, other than not having to walk from brokerage to brokerage to find out about different houses.) But the realtors seem to forget that *Why The Heck* is not proposing that brokering for commissions be made illegal.

Responding to a comment that "if the buyer had to write half the check, that half would almost disappear within days," a broker responded: "The buyer does write half the check, whether they (sic) realize it or not." Did I miss something in my own posting, or isn't that precisely the point—that buyers don't realize they are paying this half? I don't think anyone would have any problem with buyers paying half of the brokerage

---

18. However, even that can't be taken for granted. I saw a listing boasting that the property had "a view of Interstate 95" and "an extermination contract with the Waltham Pest Control Company."

commission, if buyers realized this explicitly and had other options. "If commissions were not in the price, everyone else would need to be doing the work." Likewise, if I don't pay a cleaning person to clean my house, I have to clean it myself. But that doesn't mean housecleaning should be paid for in my mortgage with no option for me to clean the house myself.

The comment continues: "It is also incorrect to assume that if a rate remains the same in a certain town it must be 'fixed.'" Okay, maybe it's just a huge coincidence that town after town has the same rate that never budges.

While I was proposing to make licensing easier, one commenter proposed making licensing harder, in order to reduce the number of brokers, which would make them busier and therefore reduce their prices. Let me see if I understand this logic: Artificially constricting the supply of vendors would reduce prices? So we should go back to having only one phone company?

Finally, even the most eloquent broker-commenter acknowledged that he would like to have at least some hourly-paid consulting clients, to smooth out his income and keep him from the trap of spending hours with prospects who never buy. Sign me up to work with that guy.

Ultimately, this is all that is being proposed here: freeing up the market to provide a range of options to people who want to use, or not use, brokers. The market would become more liquid, buyers would pay less even as the seller realized a higher net, and the number of houses going into foreclosure should drop quite considerably.

CHAPTER SEVEN

Your Medicare Tax Dollars at Work. Do You Mind If I Step Inside?

I once had a summer job selling *Collier's Encyclopedia* door to door. Collier's was celebrating its hundredth anniversary, a promotion which, I later learned, worked so well that it had been held over for seventeen years. Collier's boasted a dedicated team of salespeople. My field manager, Kip, was so profoundly committed to Education For The Home that if you declined to buy the books, choosing instead to deprive your children of the value of Education For The Home, he'd pee on your lawn.

Which is precisely my point. Kip can pee on your lawn all he wants — just not at government expense. His bathroom habits are between you, him, your landscaper, his urologist, and Collier's senior management team, who no doubt would be stroking out right now if Collier's, like most other door-to-door encyclopedia companies, hadn't already gone bankrupt.[19] Yet despite their generally low integrity and bladder capacity, *no*

---

19. Darn that Wikipedia

Collier's salesperson ever once suggested that taxpayers should pick up the tab for the entire 24-volume Home Major Reference Library. Not even for the Complete Shakespeare Collection.

And therein lies the key distinction between Collier's and Medicare, other than that the latter is not yet bankrupt. Depending on who's counting, 5% to 10% of the tax dollars you pay to the federal government for Medicare HMOs, which you assume are being used to finance health care for senior citizens, is going instead to salespeople and brokers whose job it is to convince seniors to join those HMOs. Some small portion even goes to brokers to convince people to stay in those HMOs. Yes, every year brokers get a residual fee too. The HMOs pay brokers all these fees for the same reason Collier's paid Kip his commissions — to convince people to buy something. In this case the "something," a Medicare health plan, is taxpayer-financed, meaning the commissions are also taxpayer-financed. This chapter explores the possibility that perhaps taxpayer-financed sales calls are not the most efficient way to give seniors their free entitlement.

Occasionally you see populist pundits calling on government to sweep away layers of bureaucracy, but personally I don't mind paying for overhead. Every company has overhead. The federal government is generally the poster child for overhead. Overhead, I understand. But brokerage fees?

Before we start being incensed, here is a little background on Medicare. A team of experts[20] scientifically determined

---

20. Actually it was Wilhelm Bismarck, as in Kaiser Bismarck. He picked that age "scientifically" because in those days most working-class people that age were dead. (Note to the 1% of you who have correctly sensed that something is amiss here: As I was recently reminded, Bismarck was never the "Kaiser" and his first name was Otto, not Wilhelm. Score one for my uncle, the history professor. However, he – meaning Bismarck – did indeed invent the 65-year-old retirement age. That much I got right.)

that 65 should be the age at which people could retire. That determination in Germany started a chain of events ultimately creating an entitlement providing most Americans over the age of 65, retired or not, taxpayer-financed health care. People obtain that health care mostly in one of two ways. First, there is traditional Medicare. People who turn 65 have access to any hospital or (with an optional Part B premium) any doctor, and Medicare usually pays a lot of the bill. Note those words "usually" and "a lot." There are so many holes, known as "gaps," in traditional Medicare that an entire industry has developed to charge senior citizens four figures a year to cover those gaps.

There are also many variations of non-universal-access plans for seniors offered by the private sector, for which we will use the shorthand, "Medicare HMOs." In areas where those are available, seniors can go out of their way to sign up for a Medicare HMO, which has a network of doctors and hospitals — not unlike the insurance plan they had the day before they turned 65. The Medicare HMOs cover enough so that their members don't need "gap" insurance. HMOs aren't exactly a hardship. I am in one and I don't consider myself a second-class citizen as a result. And I will certainly join a Medicare HMO too, if I manage to turn 65 without being whacked by the real estate brokerage or zinc lobbies along the way.

Like many regular HMOs including mine, Medicare HMOs also offer special features at no extra cost, like free health club memberships. Also, whatever else one argues about, Medicare HMOs should be offering somewhat better quality than traditional Medicare, because they have to track and report all sorts of clinical and service quality indicators that traditional Medicare doesn't.

They also offer "disease management." Disease management provides nurses and other resources *gratis* to help people manage their own chronic diseases so they don't end up in the hospital for preventable complications. It is a concept near and dear to my own heart because if you google on "invented disease management," you get none other than *moi*. Because disease management is health care at its quintessential best — investing in prevention to avoid needless suffering and cost — I would be very proud of having invented it if the story were even the slightest bit true[21] Whatever, you just need to know that the HMOs offer it and traditional Medicare doesn't.

So far this all makes sense in theory — a program that limits provider choice and provides benefits to keep people healthier should cost less than one that doesn't. Since Medicare spending periodically threatens to bankrupt the Government if we can't bring it under control, it would seem that an easy "fix" would be to limit members' choice to those providers willing to discount, or at least make it more expensive for members to go "out of network." After all, most of us are in health plans today that do exactly that. If the alternative is to bankrupt the system, why can't we continue to subscribe to similarly designed plans after we retire, or be willing to pay extra for the privilege of unlimited choice out of our own pockets?

This all assumes that unlimited provider choice costs more than limited provider choice. Hence a traditional Medicare plan offering unlimited choice should cost taxpayers more, right?

Wrong. The government may or may not be actually spending more money per person in Medicare HMOs than it does on traditional Medicare. It depends who you ask, what counts

---

21. Don't tell Google but the reason I know I didn't invent it is that the first time I heard the term, my reaction was "What's disease management?" I don't recall that Einstein ever asked: "What's relativity?"

as "cost" and what specific region of the country the member lives in, and, within that region, whether the member lives in a city. For *Why The Heck* purposes, let us split the difference and assume that under current circumstances the taxpayer has no preference between the two approaches. Yet the HMOs offer better benefits, more coverage, and more accountability for the quality of their providers. However, "current circumstances" can be improved in two ways to save money in Medicare HMOs. Whether the savings goes to taxpayers in the form of lower premiums for HMOs or to HMO members in the form of better benefits is a political question. *Why The Heck*, however, is about efficiencies and the rest of this chapter is about two inefficiencies that should be addressed.

Inefficiency #1: It takes extra effort to join a Medicare HMO

Medicare's regulations make it cumbersome for people to join HMOs, which in turn makes it expensive for the Medicare HMO to enroll people. People are given the traditional Medicare plan as a "default" and have to be "sold" on joining the HMO. This is where the brokerage fees come in and now is the time for you to get incensed. Because of this unlevel playing field between the HMOs and the traditional plan, Medicare HMOs need to pay brokers to get people to join their health plans. Those broker payments are coming out of our taxes. In fact, brokers who enroll senior citizens in these HMOs almost always gross more on those members than the health plans themselves make. In other words, broker fees are a multiple of profit for most plans.

A health plan can spend up to $100 on a new-member broker fee, contributing to a total member acquisition cost (including advertising, enrollment, and orientation) that I will call 5%-10% of revenues. It varies by health plan, but it is usually closer to 10% of revenues. And, just in case this isn't already abundantly clear, their "revenues" are what the Government gives them—our tax dollars. Yes, we are paying the Kips of the world to pitch HMOs to seniors.

There is a lot of debate about the cost-effectiveness for taxpayers of these HMOs generally. But however cost-effective they are today, they are up to 10% *less* cost-effective because they have to sell people on giving up traditional Medicare...even though the limited-network HMO-type structure should be a lot more familiar to people than the gap-filled universal provider traditional model. In some cases in some markets, the HMO itself could be the same as the one that had covered them the day before they turned 65.

Why these sales expenses? Why should people have to be "sold" at taxpayer expense to leave an unfamiliar and expensive plan to enter another plan that covers more, costs less, and offers more quality monitoring? Unless someone can explain exactly why this extra transactions burden on seniors and taxpayers is justified, we have one simple proposal to reduce that burden.

The simple proposal: instead of an "opt-in" Medicare HMO program that requires health plans to pay brokers to enroll members, pay the members directly to enroll themselves. Replace the current system where plans pay brokers to get people to "opt in" to an HMO with a more neutral but win-win Medicare HMO signup policy, where health plans can pay up to $100 to 65-year-olds who initially sign up, using Medicare's introductory ailing rather than a broker. Seniors who cash the check agree to stay in their HMOs for some number of months, after which they can either pick whatever plan they want, or go back to Medicare.

Here's the way it would work. Medicare would send initial mailings to newly minted 65-year-olds, just as they do today, but the mailing would include a check if they try the HMO option for a few months. They are given a list of participating HMOs to choose from, listed according to their ratings in the Medicare "Star System" for quality. Perhaps only HMOs that are willing to give up their outbound sales or brokering function can participate in the signing-bonus system. (Alternatively, they may keep their outbound sales function, but not for 65-year-olds.)

Of course, the HMOs can maintain their inbound member services function for these prospective members, so that if they want to call to find out more about the health plan, they can.

Everybody wins, except the middlemen.

It should be emphasized that people are totally welcome to ignore the mailing and the check and stay in regular Medicare.

But only as a backup. There's no reason why someone who is 64 and is perfectly satisfied in a health plan with a limited network suddenly needs to be "re-sold" on that concept, by a broker, the day he or she turns 65, especially when they (1) are getting a sign-up check and (2) save a four-figure sum that they would have spent on gap insurance if they chose regular Medicare.

One way or another, many people will select an HMO and stay with it. Today, 90% of people who join a plan renew their membership a year later. Which brings me to one final point: Just because it takes the Earth 365 days to go around the sun, people should not have to be re-sold every year, once again at taxpayer expense, to stay in those plans because the rules require that people affirmatively re-enroll, rather than passively stay in. That is indeed what happens today. Instead, if they want out, they should be able to get out. Otherwise renewal should be automatic. People can stay in their plans as long as they like. People who want to change plans or go back to traditional Medicare can take that initiative themselves, just as they do today between anniversary dates.

Inefficiency #2: People who can't afford unlimited coverage can often still get it

Now go to the next step. Suppose someone selects the universal coverage option of traditional Medicare. For the luxury of greater choice, they should pay for the "gap" insurance out of their own pockets, right? Like, if I want a Chevrolet and you agree to pay for a Chevrolet, that's fine. But if you are paying for a Chevrolet and I want a Cadillac, you'd say I had to make up the difference, right?

Well, here's a little taxpayer trivia for you: If seniors select the traditional Medicare and can't afford to pay for the gaps policy and have income below a threshold, guess who does pay for it? If you guessed "taxpayers" (this time in the form of Medicaid), good for you. You don't need Education For The Home.

So let me see if I have this straight. Instead of "settling" for an HMO that is good enough for me and millions of others and that demonstrably offers more services and more quality monitoring, senior citizens who cannot afford "Medigap" insurance may instead select the traditional gap-intensive Medicare policy and in some cases get taxpayers to cover those gaps? I find this troubling. Whoever came up with this idea, better not have a lawn.

COMMENTS

Since my "day job" is in healthcare services, I know how hard it is to propose even the slightest efficiencies without endangering someone's livelihood. Consequently, I put a version of this proposal up on www.theheathcareblog.com to attract the negative comments, as an electronic lightning rod, rather than wait for the book to be attacked. Naturally the opposition came from brokers, who said that they do a lot of work to convince people to join these plans and explain the benefits of them. No commenter disputed the premise that an "opt-out" level playing field for Medicare would make it much less expensive to entice people into HMOs. This is pure *Why The Heck*—a reduction of nonproductive healthcare expense that makes the economy more efficient.

POSITIVE COMMENTS

The positive comments were nowhere near as interesting, as passionate or as personal as the negative ones. More interesting was the occasional acknowledgment in the negative comments that I had a point. No one disputed the premise that explaining benefits is a much less expensive way to do business than sales. As a taxpayer I don't mind paying to have someone explain benefits to people. But there's a big difference between that and collecting broker fees to sell them an entitlement.

Another commenter noted that some large employers whose employees are members of commercial health plans that also offer Medicare HMOs do sometimes indeed seamlessly transition those members at age 65. This offer provides "proof of concept" for the first part of the chapter, illustrating that the concept of a network-based health plan that provides limited choice is not foreign to employees today.

It was also noted that this brokerage expense was never on the table during the health reform debate. Because it is paid for by the health plans rather than taxpayers directly, it doesn't show up as a government expense.

NEGATIVE COMMENTS

One comment from a well-known Medicare HMO expert noted that the database that would allow Medicare to set up this level playing field is still rather incomplete. Medicare can know who is turning 65 through the Social Security Administration, but not which plan, physician(s) and hospital(s) they use now. This data is "out there" but would require some coordination.

The expert also noted that the Government tried online enrollment and information when Medicare introduced its Part D drug benefit in 2006, but many people couldn't use it. That is true. Two differences, though. First, Medicare tried enrolling everyone online but some people lacked sufficient familiarity with computers. This proposal would skew towards the younger, newly eligible members, who do know how to use them. Second, even among older people, Internet literacy and access have increased since 2006 and continue to do so.

He also emphasized that there will still be the need for disseminating information by telephone, even once the playing field becomes level.

One person wrote that, in the business world, a 10% customer acquisition expense is "not bad." Perhaps true in general, but for a taxpayer-financed entitlement? I find it discomfiting that any significant amount of my taxes goes to brokers to entice people to do anything. He continues that if HMOs are that good, they should stand on their own merits. True, but aren't they entitled to try to do so on a level playing field, rather than having to spend 10% to be "sold" when the traditional Medicare enrollment is automatic, with no sales expense?

Another commenter said that Medicare HMOs cost so much more than traditional Medicare that putting more people in them raises the total cost of Medicare, sales expense or not. However, the additional expense is partly due to the fact that the HMO benefit is somewhat richer, so people don't need "gap" coverage—a savings of four figures per person. Once again, *Why The Heck* does not challenge people's data. It says, regardless of who is right on this particular question or any other, the leveling of the playing field saves money for the system as long as it has been determined that Medicare HMOs are cost-effective enough to be offered at all.

A broker wrote in to say that Medicare HMOs were hard to explain to people, that Medicare itself spends a lot of money simply explaining the basic benefits, and that's why brokers should get paid to do so. I don't doubt that brokers have to explain a lot of stuff, but there is a big dollar difference between the cost of explanation and the cost of brokering. Remember, the real estate brokerage industry has resisted allowing its brokers to charge a consulting fee for explanations and consulting for exactly that reason. However, the underlying point is well-taken: Because there is still a cost of explanation, the entire 10% member acquisition cost would not go to zero.

Also, why do the people who run Medicare seem to regard its own HMOs as "competition" and keep these plans at arm's length? Why should there be two different sets of explanations, one for traditional Medicare and then one from the broker for HMOs? That is indeed the crux of the proposal, to lay the whole thing out in front of people. Yes, explanation is necessary but why complicate it further by having it in two different settings? And people who qualify for Medicare today come from the first generation of workers who have mostly been enrolled in HMO-like health plans at work. Over time, it should become easier to explain that a Medicare HMO is like the HMO that insured them the day before they turned 65.

Three of the negative comments paradoxically made the positive case.

A broker said he often has to "go to their place" to sell Medicare HMOs to people. I don't want my tax dollars paying for door-to-door sales whether it's this guy or Kip. Another commenter said that the reason broker fees were so high was that "2-4" people had to be contacted, in order to sell one. If I don't want my taxes used for successful door-to-door sales, I am even less eager to have them used for *unsuccessful* door-to-door sales.

Another observation was that elderly people get calls from scam artists all the time. I am sure he is right. However, in an "opt-out" program, where people get the information and are told, "if we don't hear from you, you will go into this plan," all the calling is inbound. If anything, the level playing field opt-out approach would reduce the opportunity for scam artists. After all, most scam artists don't sit by the phone waiting for you to call them.

Finally, I had made the point that there is no need to pay a broker an annual renewal fee. The broker pointed out that absent an annual renewal fee, brokers "have an incentive to move their entire book of business ever year." Because the annual renewal brokerage fee is only a fraction of the initial brokerage fee, that incentive exists anyway, but eliminating the concept of sales in the first place also eliminates the need for a renewal brokerage fee.

As usual, there was at least one pony in the negative pile. Someone asked, in a city with multiple HMOs, which one would be assigned as a default? We had proposed some coordination with the private sector, but this is a "second-order problem" that the health plan industry should be able to figure out on their own through their trade organization.

This chapter proposes that the commercial insurance plan would have to send its data to Medicare so that Medicare could match up the providers a member has been visiting with the providers in the network of the various HMOs, and assign the one with the best apparent match as the "default" plan. There are other possibilities:

The HMOs could bid for the right of being the default, or alternate "first position."

Health reform already puts some teeth in the HMO quality measurements mentioned earlier, rewarding "5-Star" Medicare health plans. Why not give them preferred access to new members as well?

And don't rule out the government's ultimate tool for decision-making and dispute resolution, rock-paper-scissors.

Part Four

Ideas Guaranteed to Attract Opposition from People Who Don't Even Know Any Middlemen

Occasionally in government, it becomes necessary to stop campaigning, fundraising for the next campaign, campaigning for one's friends, and fundraising for their next campaigns, and instead actually govern. "To govern" can be defined as: "to take the steps necessary to ensure the functioning of a society using the Rule of Law, under a duly ratified Constitution, including the implementation of the ideas in *Why The Heck*."

For all I know, there may be other definitions as well but, no matter how one defines it, the solutions in Part Four are going to require it. There could be actual opposition from people other than middlemen whose livelihoods are at stake. For this next set of ideas, instead of being based on narrow-minded self-interest, the opposition would be based on deeply held principles. It would come from people whose patriotism cannot be questioned just because of a philosophical difference of opinion, whose motives are pure if perhaps misguided, and whose IQs are about as high as Bart Simpson's [22]

Chapter Eight—we can't put it off any longer—tackles actual health care delivery and offers simple, incentive-based ways to save a few percentage points of spending that was apparently too controversial for the actual health reform law because no one had proposed a "kinder, gentler death panel" to replace the original death panel when it was booed off the stage early in the reform debate. Even though my death panel proposals are kinder and gentler than the original one (which "threatened" to reimburse physicians for talking to their patients about end-of-life issues), this will be the first idea in the book to get a few people other than middlepeople really annoyed with me, but get used to it—there will be more in the three subsequent chapters. The emphasis, though, is on the "few." Even as the ideas in this book become progressively more stimulating in both senses of the word, they would always benefit the vast majority of people.

In Chapter Nine, Toto pulls the curtain away from me. You bought this book thinking it was all original thinking, but that turns out to be a lie. The harsh reality is that no single component of Chapter Nine is original, only the sum total. This chapter points out the sustainability of the American recovery depends on fuel prices not because they are too high, but rather because they aren't high enough. Fuel prices, especially gasoline, are way too low and not for the reason environmentalists tell you, which is that they don't include the cost of global warming.

---

22. And I am not referring to the episode where, after Homer and Marge put Bart on Focusin to raise his IQ, Bart announces that "the average person only uses 10% of their brain, and now I'm one of them."

No. Gas is too cheap because basic principles of cost accounting would say that the dollars being spent on defending our oil supply lines should be allocated to the product on whose behalf they are incurred. As a result, gasoline is underpriced and overconsumed. Few people realize the damage that too-cheap gas (and too-cheap energy generally) does to this country and to the recovery. To fix this situation, we propose the "Green Dividend," a relatively painless way of bringing gasoline prices up to a level that most accountants or economists would agree covers those barely hidden costs. A Green Dividend compensates people for increased gas taxes fully, in advance, and visibly.

Chapter Ten takes another, longer look at the cost of being poor, and how to reduce it even more than the Instant Tax Refund Anticipation Loan checkoff box. Once again, most of you won't have any idea what I am talking about here, and frankly even I had to do actual research. Not "research" a la Barbara Ehrenreich where I actually subject myself to the indignities of poverty, but "research" as in checking in with my friend John Hoffmire, who claims to know poor people. [23]

Chapter Eleven proposes a novel way to raise demand for housing and increase consumer spending while providing long-term Methadone-free economic stimulus. It turns out that one way to wean ourselves off the Methadone Economy is right on America's doorstep: Immigrants. No, this is NOT about illegal aliens and border walls and amnesty and Jerry forgetting to

---

23. In all seriousness, John does more than claim to know poor people. He is the Director of the Center on Business and Poverty (www.cobap.org). I support it and would recommend it to others. He and his organization inspired this chapter and also taught me about tax refund anticipation loans. His organization helps some corporations do, on a private and voluntary level, much of what is proposed in this chapter.

give Babu his visa renewal application and a whole bunch of other things I know nothing about and really don't even have much of an opinion on. It's pure Economics 101: Bring in well-heeled immigrants with money to spend and they will live in our unsold houses and buy our unsold goods, while setting up businesses that will employ our unemployed workers.

CHAPTER EIGHT

A Kinder, Gentler Death Panel

This chapter is on death and dying. Death and dying is not funny. Even the episode where Susan died from licking George's cheap wedding envelopes wasn't as funny as, for example, the Soup Nazi. So this chapter will not sound like the rest of *Why The Heck.* Instead it will sound like it was written by an actual adult.

Let's tiptoe into this topic with an analogy. Imagine a restaurant that insisted on giving you more food than you wanted, and wouldn't let you leave until you ate it all. If you wanted less, you had to make that request known well in advance and if the waiter didn't like the way you asked for less, he'd give you more food and make you eat it all anyway. (In that sense he would be the opposite of the Soup Nazi.)

Such a restaurant would run out of food faster than Lehman Brothers ran out of mortgage insurance, except that, unlike Lehman Brothers, which had to pay its own bills, in this restaurant the waiter would send the other diners the check for your extra portions, and they would pay every time—sort of an ongoing bailout for overstuffed eaters.

Preposterous? Of course. Yet that is exactly how dying works. In states other than Oregon and Washington, which have right-to-die laws, you are not allowed to hasten your own deaths no matter how ill you are. You have to wait for nature to take its course. And that's the best-case scenario. The worst-case scenario is that you are willing to wait for nature to take its course but are accidentally or inadvertently kept alive too long, often against your wishes. If that happens, today it's usually your own stupid fault because you failed to communicate your wishes according to the exacting specifications required to do so. Or if you did communicate them, the hospital can't find the little piece of paper the communication was written on.

Without that communication, your inadequately expressed desire for comfort care gets trumped by hospital financial incentives, which favor more taxpayer-financed medical and surgical interventioThe reason this issue makes it into *Why The Heck* is quite simple: Dying is expensive and inefficient—even as compared to the rest of our healthcare system, which is quite a feat. One of the tenets of *Why The Heck* is to reduce "drags" on the economy, of which health care—particularly government-financed health care—is high on the list, reformed or not. Medicare spends one-third of its entire budget on the final two years of life for chronically ill people, years fraught with ultimately futile and usually debilitating and painful medical care. [24]

---

24. Unless otherwise noted, the statistics on utilization and cost in this chapter are drawn from *The Dartmouth Atlas of Health Care* (www.dartmouthatlas.org), published by the Dartmouth Institute for Health Policy and Clinical Practice. [back] However, the conclusions are mine so I am the guy to holler at if you don't like them.

Why The Heck doesn't *Why The Heck* talk about Health Care Reform Generally?

Hey, Einstein, have you paid any attention at all to the six chapters of this book that preceded this one? Enough attention to notice that there have actually been seven chapters preceding this one? If so, surely you've noticed by now that *Why The Heck* doesn't address government initiatives that require being financed by someone ("someone being defined as "you and me"). We address only win-wins. We were neutral one health care reform itself. Personally, though, I'm glad it passed, if only because my COBRA is going to run out in a few months. (Yes, I know. As my first wife used to say, it's not always about me.)

Admittedly, that observation carries an asterisk, which is that for many conditions no one knows in advance the specific date on which that final year begins. Unlike Super Sunday, the day your taxes are due, or the day you forget your anniversary, you can't just mark it in your calendar. If you knew exactly when you were going to die, you would do a lot of things differently. You would get your affairs in order. You would forgive any past grievances with family members because they are, after all, your flesh-and-blood and deep down you love one another, you share emotional and physical bonds going back in some cases almost a century, and what better time to reconcile than now when emotional support is so profoundly needed? Plus you would watch all the shows you've TiVoed.

Even with that asterisk, keeping terminally ill people alive against their will, the epitome of cost-ineffectiveness, is a "tradition" whose time has passed. Once again, this is *Why The Heck*, which means that everything is voluntary. *Why The Heck* is not:

- saying that people can't be kept alive if they want to be;
- rationing end-of-life care; or
- denying any access to treatments.

Surprisingly, one does not need any of those mandates to reduce costs. Costs can be reduced quite a bit simply by improving the dying experience — adhering more closely to the wishes of patients, and aligning incentives between doctors, hospitals, and patients.

Why The Heck will propose seven ways to do this, but first, let's provide a little primer on the topic of dying, and introduce you to the concepts of advance directives, palliative care/hospice, and give you a sense of how hospitals get paid today.

Advance directives

An individual uses an advance directive to specify, in advance, treatment options for contingencies in which he or she would not be capable of making a decision on the spot. An example of such a contingency would be permanent unconsciousness, as might result from a massive stroke or a prolonged period of oxygen deprivation to the brain following resuscitation from cardiac arrest. The phrase "permanent unconsciousness" includes both coma, in which a sleeping patient cannot be awakened, and vegetative state, in which a patient, despite having periods of "wakefulness" (eyes open, movement) shows no awareness of his or her surroundings, and is therefore "unconscious" of the environment. In an advance directive, an individual specifies whether or not, in the case of permanent loss of consciousness, he or she wants to be kept alive by artificial means, such as a ventilator or a feeding tube.

Advance directives come in different forms in different states. A living will specifies a patient's wishes in different circumstances, while a health care proxy (or power of attorney for health care) designates a trusted friend or family member to make medical decisions if a patient cannot do so. Even with both documents in place, contingencies may arise that no one thought of, or are not thoroughly covered, or the "directive" may be ambiguous. Other times, physicians may disagree on whether a patient is indeed terminally ill. Imperfect as they are, advance directives are nonetheless a useful means of assuring that a patient gets only the medical interventions he or she really wants.

And what do people really want? One interesting observation about coverage for permanent unconsciousness comes from a *USA Today/Kaiser/ABC* poll asking people how they feel about artificial life extension. According to this poll, 40% of us want to keep people alive, and presumably be kept alive ourselves, at all costs.[25] But costs to whom? Taxpayers, that's who(m). The expense is invisible to the actual patient. So of course some of us want to be kept alive at all costs: It's free. Under that scenario many people would want Botox too. One can only guess what percentage of people would choose the "keep alive at all costs" option if they were paying for it.

What is much more guessable, though, is how many people in actual practice want to be kept alive when they are told that with virtual certainty there is no chance of regaining anything close to health. To get into that statistic, we need a brief primer on palliative care.

Palliative care

By way of background, many people who are terminally ill when admitted to the hospital have not been counseled by their own doctor as to what to expect, and many lack advance directives as well. Even those who do have advance directives and have been counseled by their doctors may be frightened

---

25. 26: Appleby, Julie, "Debate Surrounds End-of-Life Healthcare Costs," *USA Today* (October 19, 2006), p. 1

and conflicted about what they want as the end approaches. In the absence of a palliative care team, these patients often find their way to the ICU, because specialists whose training and remuneration are both geared to doing more interventions are not as comfortable or as well-paid for doing fewer, or even for discussing realistic alternatives to critical care. Not to mention that those ICUs need to be filled. (More on that later.)

The palliative care physician's job is just the opposite—to counsel terminally ill people on all their options, from intensive care to comfort-only care. The palliative care physician has no financial stake in the patient's choice. Pretty consistently, two-thirds elect comfort-only care with perhaps only minor interventions like antibiotics. The *USA Today* poll proves not to be far off in practice: About a third of people who are counseled regarding all alternatives choose the ICU, ventilator, and resuscitation in the event of cardiac arrest—the whole ball of taxpayer-financed wax. However, two-thirds choose to limit aggressive care, preferring to die with some semblance of dignity and comfort, perhaps in a home hospice setting.

Either way, the palliative care team (nurses and social workers as well as physicians) helps terminally ill people avoid futile care if they don't want it but still choose it if they do. The team also addresses the fact that most terminally ill people don't know what to expect in the hospital, or helps them when they suddenly change their mind.

Payments by Medicare to hospitals and doctors

Years ago, some government contracts were procured on a "cost-plus" basis. The defense contractors' profit was based on their costs, so the more they spent, the more they made. Needless to say, the media watchdogs jumped on that one and as a result the cost-plus practice has fallen by the wayside as a method for government payments to suppliers. Except in hospitals, where in some cases the more the hospital does to a patient, the more they get paid, using the "diagnosis-related group," or DRG system. Hospitals tell Medicare what they did to a patient and what diagnosis a patient has, and Medicare reimburses them off a schedule based on those two items. The more they do to a patient up to a point, the more they get paid.

I will let someone else do the talking here. Joseph Sacco, M.D., who specializes in palliative medicine at Bronx-Lebanon Hospital Center and who also teaches the subject, recounts some of the difficulties he has encountered in providing appropriate end-of-life care:

> A hospital is a business. Like any businessman, a smart hospital CFO will do what he can to maximize revenues. In health care, that means taking aggressive care of terminally ill patients. Such situations constitute one of the highest-revenue non-transplant patient categories in all of Medicare, a category known as "DRG 542." A patient who spends five or more days on a ventilator and undergoes a tracheostomy and a major surgical procedure falls under DRG 542. The hospital's reimbursement from Medicare for DRG 542 often reaches six figures.

To maximize revenues for patients on ventilators, the hospital needs to perform a tracheostomy [26] after the patient has been on a ventilator for at least five days, but before "pulling the plug." Pulling the plug before the tracheostomy means

---

26. For those of you who either think my spellchecker isn't working or else want to know how a tracheostomy differs from a tracheotomy, here is your answer: A tracheostomy is a surgical procedure. An "otomy" is medical jargon for any hole, meaning that a tracheotomy is what happens when a friend of yours is choking in a restaurant and the Heimlich doesn't work, and you ask if anyone has a pen handy

reduced reimbursement. However, to maximize profits the patient shouldn't be lying around on a ventilator *without* a tracheostomy either. The Intensive Care Unit (ICU) days are adding up, but reimbursement is not increasing.

For patients with terminal illness such as late stage heart and lung disease, cancer or cirrhosis, five days on a ventilator followed by a tracheostomy and a major surgical procedure is unlikely to do anything other than prolong dying. At best, these patients are extremely unlikely to leave the hospital in any kind of a functional state, and will become critically ill again in weeks if not days. Yet they are also the most profitable patients. The hospital gets paid the most not for restoring people to health but for keeping nearly dead people alive. I know what you're thinking, "Al, you can't be just about to say that hospitals and doctors would try to fill beds with the highest-paying patients just to get paid to keep them alive. Financial considerations would never trump clinical integrity. That would lead to the ridiculous conclusion that the more hospital beds there are, the more people will be placed in them. Everyone knows that people are only admitted to the hospital when they really need to be."

If you're thinking that's what I'm just about to say, you'd be wrong. I am not "just about to say" it. An actual authoritative source is just about to say it, and all I am going to do is cite that source. According to *The Dartmouth Atlas*, which compiles and compares medical statistics by region, cities that have more hospital beds to fill and more doctors to feed tend to have more days of intensive care. The cost of your final year would be almost twice as high in Miami, Florida, as in Portland, Oregon. In Miami, in the last six months of life alone, you'd have 46 doctor appointments, mostly with specialists. In Portland, you'd have only 18, and half of those would be with your primary care doctor.

The same is true for days in the ICU: In that final year, Miami residents have six ICU days, Portland residents only one. The primary reason for this disparity is, according to this very same *The Dartmouth Atlas*, that Miami has more ICU beds and specialists than most other places. Because humans have the same physiology in Portland as in Miami and because life expectancies are similar in both cities, it is unlikely that the difference in utilization of ICU beds and specialists is caused by anything other than the conclusion reached by the *Dartmouth Atlas* using many examples from many cities: More ICU beds and more specialists generate more ICU days and more specialist visits.

So it turns out that hospitals and physicians, like real estate brokers and other people such as you and me, tend to be driven by economics, in this case the economics of one of Medicare's highest reimbursement categories, DRG 542. Change the economics to align incentives, and you'll change behavior. Futile care is like anything else: Pay for less, and you'll get less. Although, once again, let us be clear: *Why The Heck* is not advocating the rationing of any form of medical care, end-of-life or otherwise. *Why The Heck* doesn't do mandates.

Doctors get reimbursed in an even more primitive manner: The more they do to a patient and the longer they have a patient in the hospital, the more they get paid.

The intersection of those three items

Viewing these three items together, one might ask whether the third is somehow in conflict with the other two. You have patients who by and large want to be left alone, managed by doctors and hospitals who would make much more money by doing exactly the opposite. Now let's get to the good stuff: what to do about it. Amazingly, at least four of these seven proposals will not need Congressional approval.

1. See how hospitals handle dying patients

Tally DRG 542 (and other DRGs and other indicators) by hospital to see which hospitals have the most aggressive treatment and then release those findings to the media

All the data is already available for this analysis. It just remains for someone to do it. Start with people who have an end-stage diagnosis of some kind, in order to cull out those people who really do have a chance to regaining a semblance of health. Within that terminally ill group, compare the number of people with DRG 542 who die in the hospital to the number people with DRG 542 who leave alive. Then, compare DRG 542 deaths to total ICU beds to see if hospitals are filling their ICU beds with dying patients, and maybe also compare the ratio of DRG 542 cases to DRG 541 (a much less aggressive cousin of DRG 542), to see if hospitals are doing more in order to make more money.

No doubt a health services researcher would come up with other comparisons too. Bottom line: A list of the "best and worst hospitals to die in" could be compiled from ingredients that the Government already has in its kitchen. Quite literally, this analysis could be done tomorrow.

Instead of a Congressional mandate, this approach shines a light on the data and then lets the Court of Public Opinion decide what to do about the findings. My suspicion is that no hospital administrator would want to be on the "worst hospitals to die in" list, and that Medicare HMOs would have a thing or two to say about it as well. There would be a scramble to improve the dying experience, which – without any emphasis on cost savings – would save money just by making people better off.

2. *Do something similar for physicians*

A very simple step would be to tally the percentage of a physician's patients who had advance directives in place at the time of death (meaning, of the patients). Then make this information available to researchers and the general public. Doctors hate to be "outliers" in quality rankings. When see their own rankings and know that patients are seeing them too, they will improve. Once again, we would rely upon the Court of Public Opinion to publicize these findings.

Unlike the first proposal, however, the data does not already exist. Someone would have to collect it. Even so, Congressional approval is not needed to collect data, so this can all be done now.

3. *Collect specific quality-of-death data for hospitals*

Hospitals are required to report many variables related to the quality of care, such as infection rates, readmission rates, and death rates in certain types of surgery. None of those variables relates to how well they handle dying patients. Perhaps certain variables that correlate very closely with quality-of-death should be brought out into the open. How many people died after more than two weeks in the ICU? How many died after one, two, or more cardiac resuscitations? What is the hospice referral rate? How many people who died had advance directives in place? How many people had multiple resuscitations from cardiac arrests? What were the satisfaction scores for the families of patients who died in the ICU? How many people on ventilators had tracheostomies and subsequently died in the hospital?

This would be a slightly more intensive data collection project but even so, it's unlikely that this type of data collection would require Congressional consideration.

4. *Make dying less profitable for the hospital*

So far all we have proposed is to collect and publish some data and embarrass hospitals and doctors into doing a better job. But now let's see if those hospital economics can be tweaked in typical *Why The Heck* fashion to produce a better outcome for patients.

Aligning financial incentives with the actual wishes of patients and their families should make more facilities more conscientious in responding to those desires. Of course doctors and hospital administrators will all be as horrified to hear that observation as real estate brokers were when I posted my essay on www.Whytheheck.com about their commission-based incentives. Medical and hospital staff would say that they already respond to patient wishes and would never put their own revenues ahead of ethics. In that case they shouldn't mind this proposal at all. I'm sure every single one of them is making best efforts already to find those advance directives, and never pushes their patients to undergo heroic one-in-a-thousand-chance-of-cure treatments instead. That's why the following joke is not the slightest bit funny:

> *Q:* Why do coffins have nails?

> *A:* To keep oncologists out.

We noted in the real estate chapter that if realtors genuinely thought they were offering clients a valuable service, they wouldn't mind having their fees become market-driven. The same is true of this advance directives proposal. In both cases, the fact that there is opposition makes my point better than I can.

It may be as simple as, for terminally ill patients, reimbursing at *one* case-averaged rate (DRG rate) for both deaths in ICU and hospice transfers. So no matter how long someone lingers in the ICU getting resuscitated before being given comfort-only care or moved to hospice, the hospital gets paid the same rate. That will align incentives. Each futile resuscitation would take thousands off the bottom line, instead of potentially adding thousands by getting the patient into a higher-revenue DRG. Futile resuscitations would soon be relegated to the medical history books, right after the chapter on leeches.

5. *Encourage hospitals to focus on helping people die*

Once incentives change, one habit certain to change is the general lack of hospital interest in helping terminally ill people die comfortably. Most hospitals don't have palliative medicine physicians on staff to do that. While births outnumber deaths in the U.S. only by about 1.7 to 1, obstetricians outnumber palliative medicine specialists by about 15 to 1, despite the much greater medical complexity and counseling needs of most dying patients as compared to most pregnant patients.

Not surprisingly given the amount of money spent on end-of-life medical care and given that palliative care is the only specialty whose goal is to provide *less* aggressive in many cases rather than more, the cost-effectiveness of palliative care could be the highest of any specialty. Yet only about half of hospitals offer palliative care at all, and as the statistic above suggests, even they may not offer enough of it.

Once again, I will let the palliative care specialist, Joe Sacco, tell the story:

> My palliative care team has had more than 600 instances in which patients signed Do Not Resuscitate (DNR) orders before suffering in-hospital cardiac arrest. All were patients with chronic and progressive illnesses who, had they been resuscitated, would have spent a few days or weeks unconscious on life support before their hearts stopped for good. That's 600 pointless cardiac arrest resuscitations avoided, 600 families spared watching their loved ones rot away in the ICU, not to mention maybe 3000 or 4000 needless additional days in the ICU. We showed that we saved the hospital $3 million a year.

What creates the *Why The Heck* opportunity is that $3 million in cost savings often if not usually translates into lost Medicare revenues for the hospital of much greater than $3 million.

Most hospitals would not emphasize a specialty that costs them money. So, absent any appetite for changing the economics, perhaps the government should do something else to encourage hospitals to hire more of these doctors, like publish a list of hospitals that have them and encourage Medicare beneficiaries to use those hospitals.

6. *Change the "default" provision if no advance directive is signed*

Perhaps the proposal with the greatest savings of all would be to change the "default" provision for end-of-life care.

In any other aspect of medicine, a patient wanting the more aggressive treatment option—like, consenting to surgery— has to fill out forms and get them witnessed. Yet in end-of-life care and only end-of-life care, patients automatically get *more* aggressive care unless they fill out forms and have them witnessed. Why not simply make it the other way around? Make it so you have to "opt in" to aggressive care, rather than "opt out"?

Remember, only 40% of people want to be kept alive at all costs (and that is before the alternatives are explained, which in at least Dr. Sacco's hospital reduces the proportion to 33%). So why do we assume that unless there is an advance directive, everyone wants to be kept alive as long as possible? Protocols for most medical treatments are so widely followed and well-established that they are used as a defense for malpractice claims today. The law reads something like this: If a plaintiff (patient) alleges that care was substandard and a physician can show that the care was provided according to a published standard, the burden falls on the plaintiff to prove substandard care. If the care was not provided according to the published standard, the burden falls on the physician to show a medically sound reason for the deviation.

Likewise, why can't there be a default protocol in the event of terminal illness? Patients could deviate from the protocol with their own advance directive but otherwise it would be the job of the physicians to do exactly what they do in all other areas of medicine, which is to apply the standard of care unless they have a good reason to deviate from that standard. Given that only a minority of patients wants to be kept alive at all costs even when they aren't paying for it, and given the huge expense to the system in end-of-life care, and given the atrocious quality of life for terminally ill people kept alive by artificial means, it is likely that the task force assigned to develop that protocol would agree on a default standard of care that would look much more like a typical completed advance directive rather than a typical experience today.

7. Create lower-cost insurance plans that don't cover long-term life support in the event of brain death

What do Botox and artificial life support for the terminally ill have in common?

Answer: They are both administered in a health care setting, but neither is health care. Both are personal choices. No one would say that I shouldn't be able to have Botox if I want it and am willing to pay for it. Why not say the same thing about artificial life support for the terminally ill? Why does Medicare, often combined with Medicaid when Medicare runs out for people who qualify for Medicaid, cover artificial life extension? Both are personal choices, not health care. Botox isn't covered. Why should life in permanent unconsciousness be covered?

Why should taxpayers be footing the bill for artificial life extension? Health care is subsidized both because society has an interest in having a healthy population and because there is an overwhelming consensus that it is not right to deny people basic health care—health care that makes life productive and pain-free—because they can't afford it. Artificial life extension doesn't fit either rationale any more than Botox does.

Let's consider the case of Terry Schiavo. The country divided into two camps: those who thought she should have been kept alive and those who thought she should be permitted to die. But was there a third way, a way that would prevent similar cases in the future from ending up in a controversial situation? Yes—Exclude life support for the terminally ill/ permanently unconscious from the basic insurance benefit and allow people to purchase such coverage as an optional rider. Taking life support out of the cost of insurance reduces the cost of insurance for people who don't want to be kept alive at all costs, while not denying that right to those who do.

If this rider had been available, Ms. Schiavo's wishes would have been known simply by noting whether she had purchased it. If she couldn't afford that rider, it would be her problem—just like it would be her problem if she wanted Botox and couldn't afford it. Neither of society's interests in tax-subsidized health care is furthered in either case.

Without the rider, maybe the basic coverage ends at the point at which, for example, two physicians not involved with the patient or family certify the extent of permanent injury to the brain. The odds of recovering from a coma or persistent vegetative state become smaller and smaller the longer it persists, and become a virtual impossibility after several months, especially for non-traumatic brain injury. For example, a guy who goes into a coma after striking his head in a car accident is more likely to recover than if he goes into a coma after his heart stops for several minutes. Still, at some point *Why The Heck* becomes medical ethics...and I know enough about the latter to know that it is time to shut up.

Those 40% of people who say they want to be kept alive themselves when there is essentially no hope of leaving the hospital or regaining consciousness could simply elect the extra rider. (Of course, they could change their coverage at an anniversary date, just as with any other insurance policy.) Prediction: Many of those 40% wouldn't want to be kept alive "at all costs" if it is their own nickel. Being kept alive may not be such a high a priority for them that they would be willing to actually buy insurance to cover the expense.

Speaking of professional judgment, it is my professional opinion that if I make any more suggestions that might possibly reduce people's lifespans by a few days or weeks, my carefully cultivated Colbert-like façade of political inscrutability might be crushed by a stampede of pro-lifers who will cut to the front of the line — currently comprised of brokers, oil executives, and zinc bosses — of those who would like to hasten my own trip to a hospice or perhaps shortcut it altogether. Hence I am going to stop here and throw the floor open for discussion.

COMMENTS

Comments on this posting, a version of which also appeared on www.thehealthcareblog.com, were almost all positive because on that blog I added a personal story of having lost my second wife to Lou Gehrig's Disease. It's tough to comment negatively on a posting where the poster lost a spouse, so it wasn't a fair test-of-concept. Even so, there are always the outliers who think it is murder to discontinue life support and were kind enough to console me with that observation about the circumstances of my own loss. Those people are welcome to apply their own beliefs to themselves and their own families (assuming it is OK with the latter) under any of the proposals above.

Also, I specifically asked people not to comment that they knew someone who was terminally ill and was put on artificial life support and given a week to live but later left the hospital and lived 20 more years, winning two marathons and a Senate seat. There are always rare exceptions to any rule. Fatal mistakes and near-misses will always be a part of health care.

Someone noted that Oregon has both the lowest costs and, until Washington joined it in 2008, the only right-to-die laws. She asked why I wasn't advocating right-to-die. Right-to-die is not an original idea, that's why. *Why The Heck* is about original ideas.

Someone else suggested a national registry for advance directives, which hospitals could call upon in the event of emergency, together with incentives for either the patient or the physician or both, for getting an advance directive into the registry. You'd get much higher compliance and no more hospital personnel searching around trying to find the thing when they suddenly realize they need it. Perhaps an advance directive registry will be part of the official Electronic Medical Record project, which is in the Stimulus plan.

NEGATIVE COMMENTS

The negative comments (aside from the person who accused me of homicide) didn't take issue with the concept but rather with the line-drawing between what constitutes artificial life support for the terminally ill and what constitutes keeping someone alive in hope of recovery. Someone said the proposal was very "Big Brother-like" because it would require someone to determine when to disconnect life support, but I think they have it backwards. It's "Big Brother-like" to make that determination in the absence of any data to the contrary. All of these proposals increase either the amount of data on hand, or the likelihood that with that data on hand, hospitals and physicians will make the choice most in accordance with patient wishes.

Another noted that Terry Schiavo may not have been able to afford the permanent-unconsciousness rider, so her wishes could not be automatically inferred from her not having it. That is true, but if that rider was that important to her, she would have found a way to afford it.

CHAPTER NINE

A Kinder, Gentler Obscenely High Gasoline Tax

This chapter proposes a gas tax DON'T HIT ME.

To make matters worse, you are about to read a chapter 100% guaranteed to contain no original thoughts. Contrary to what you were led to believe when you plunked down good money or bad Stimulus Stamps for this book, every substantive idea in this chapter has been thought of before. However, here are three reasons to overlook this apparent lack of originality:

1. No one has ever put all the pieces of a gasoline tax together before into a "Green Dividend." Putting stuff together counts too, and combinations are not always self-evident. For instance, while peanut butter was invented around 1890 and the chocolate bar was invented around 1920, Reese Cups weren't invented until 1963. Yes, it took 43 years before someone eating a chocolate bar fell into a manhole right on top of someone who was eating peanut butter;

2. Now, imagine that one of the "inventions" being combined didn't get any "play" when it was first invented, sort of like penicillin. Penicillin was invented in 1929 but sat on the shelf for 13 years until two other people publicized it...and as a result of their contribution, shared the Nobel Prize with Alexander Fleming because sometimes providing the "play" is as important as doing the inventing;

3. Would you buy a product called Celluwipes? No? That's what you think. You buy it all the time, except that it got renamed Kleenex, so that chumps like us would plunk down $3 for processed wood scrapings and think we were getting a deal. Bottom line: Branding counts too, and until now, no one has come up with a way to make the Celluwipe of a gas tax sound like the Kleenex of a Green Dividend. Branding is *huge*. Even Thomas Friedman, whose books are generally speaking much better than mine,[27] missed the importance of this one. (He does actually propose branding a gas tax, but as a "Patriot Tax." I, on the other hand, don't think one can brand anything with the word "Tax" in it.)

I hope that these three observations can convince you to cut me some slack on the lack of originality for the chapter. But save a little slack to cut me on the actual tax itself. Don't rush to judgment until you read about how one can apply the *Why The Heck* philosophy to a gas tax so as to make it sound not just palatable but desirable. Start by considering three observations in general that should be self-evident, though you may at first wonder what they have to do with gasoline.

---

27. But my book costs less. Also, because I use shorter words, mine emits less CO2. [

1. Cost Accounting 101 says that resource costs should be allocated to the price of the products that use them. For instance, when you buy a house, the builder's price includes the plumbing, wood, doorknobs, and so on, plus his time and labor, and of course timely interest payments on the construction loan (ha-ha, good one, Al).

2. Uncertainty is a cost. That's why we buy insurance, to avoid uncertainties. We pay someone to take that risk for us. Likewise, people happily "pay" to achieve certainty in their investing. That's why, over time, there is less profit potential from 90-day treasury bills than from a well-diversified portfolio of professionally selected stocks (ha-ha, good one, Al).

3. It is easier to compete against a non-subsidized alternative than against an alternative whose market price includes hidden subsidies. For instance, China allegedly undervalues its currency to make its exports appear cheaper, and consistently fails to make pollution control investments that might raise the costs of their production. If China did not subsidize its products in these two ways, if the international trade playing field were truly level, the United States would still be exporting textiles and manufactured goods to China instead of the reverse (ha-ha, good one, Al).

If you do NOT believe the above—if you think products should be priced randomly rather than based on their costs, if you hate certainty so much that you prefer not to know where your kids are but would rather just guess, and if you think it's easy to sell products against competitors who give them away—you can stop reading right here. And so can your imaginary friend.

For the rest of you, let's apply each of these items to fossil fuels, gasoline in particular.

1. Cost Accounting 101 says that resource costs should be allocated to the price of the products that use them.

There are at least four "resources" that are part of the cost of gasoline (and, in two cases, other fossil fuels) but are not covered in the price of those products. The first, carbon dioxide emissions, is still somewhat controversial. Though I suspect the majority of *Why The Heck* readers believe that global warming is "real," the philosophy of *Why The Heck* is intended to appeal to almost everybody, based on logic that is almost impossible to dispute. Global warming is not based on logic. It's based on science and anything based on science will always have its detractors. Until they were faced with jail sentences, for instance, tobacco executives disputed findings that smoking is hazardous to one's health. And don't get me started on Intelligent Design.

Therefore, it is important to stick to two indisputable "external" costs of gasoline not reflected in the price: (1) traffic/pollution/street use, and (2) the cost of ensuring a steady supply of oil imports no matter what the political situation in the world. The first has been done to death. The insight that one person driving a car increases the burden on others is not original. Driving pollutes the air, and adding more cars to the road increases the time it takes any given car to reach its destination. City streets are financed out of property taxes. So every time I drive on them, you pay for the trip. And don't think I'm not grateful.

Also, to paraphrase Milton Friedman, there is no such thing as free parking.

Instead of rehashing that first set of obvious external costs, *Why The Heck* focuses on the second insight, analogous to penicillin in that has been "invented" but that almost no one has noticed it: A plurality of our defense budget is allocated to defending our oil supply lines from threats real and imagined. Oil comes mostly from unfriendly countries and allegedly very friendly and stable countries that also happen—surely by sheer coincidence—to be the home countries of most of the 9/11 terrorists.[28]

Put another way, if all of our oil supplies could be produced domestically, the defense budget could be much lower. Many would argue we would not have fought a war in Iraq. The argument that Iraq was supposed to start paying us back for the war out of future oil revenues was one of the justifications for going in. It would have been much harder to marshal public support for the Iraq War without that argument. Unlike the other arguments used to justify the Iraq War, this particular argument could not be disputed with actual facts simply because the actual facts hadn't yet had the chance to not happen.

Why should the portion of the defense budget that directly or indirectly defends gasoline supply lines be financed out of general revenues instead of gasoline sales? Why shouldn't the income tax and other taxes be lower and the tax on gasoline be higher? Please do NOT say, "Because defense is a government activity for the public good and should be paid out of public revenues." Increasing dependence on carbon-based fuels imported from unfriendly countries is NOT a public good. By contrast, research is a public good, and yet space shuttle

---

28. See, for example, Michael T. Klare's *Blood and Oil: The Dangers and Consequences of America's Growing Dependency on Imported Petroleum (American Empire Project)*. See also the report at The International Center for Technology Assessment, www.icta.org

research projects are paid for by the companies that sponsor them, not through general tax revenues.

You might say: "What does it matter whether we pay for it one way or the other? We are still paying for it. And most of us drive, so that what we would save in lowering other taxes we would simply give back in higher gasoline taxes."

If (1) you've read this far into *Why The Heck* for any reason other than to catch the cultural allusions, or (2) you've lived on this planet for a while, or (3) your Day Job does not involve setting reimbursement rates for Medicare, you have no doubt observed that the amount of a product consumed correlates inversely with its price. Raising the price of gasoline to capture its full cost would cause most people to drive less.

Less gasoline consumption is not just about less traffic, less pollution, and less financial support for dictatorships. Artificially low pricing for gasoline has created a spider-that-swallowed-the-fly effect that has rippled throughout the U.S. and world economies in many insidious ways. Here is just one example of what happened because the market price of gasoline has included so much indirect subsidization: The government has ended up subsidizing alternatives as well. Biofuels would be a perfect example. They got direct subsidies, in order to be price-competitive with indirectly subsidized fossil fuels. Massive amounts of farmland started going into biofuels production instead of food, and the ensuing shortage of food caused prices to rise for staples on which much of the world depended. And in Indonesia, people burnt down rainforests to grow corn to sell for ethanol.

2. *Uncertainty is a cost.*

Knowing that the price of anything is going to remain at a certain level or rise at a specified, predictable pace facilitates decision-making about the future. For instance, if you had known that housing prices were going to plummet, you would have rented or lived with your Uncle Max instead of buying a house. We usually assume that, in the immortal words of the great philosopher Doris Day, the future's not ours to see. But what if Ms. Day were wrong? What if the price of gasoline were to follow a specific, knowable trend or be at a certain level for the foreseeable future, and everyone knew what that future level would be? Think of all the good things that would happen:

- Can anybody spell hybrids and electric cars? Auto makers could focus on making cars whose gas mileage reflects the known future cost of fuel. This kind of certainty would allow them to build those new models on a scale that reduces their prices much faster than any government subsidy

- Buyers would buy those cars and, unlike during the summer of 2008 when gasoline prices spiked, plenty of those models would be available because the price of fuel had been correctly predicted.

- Companies trying to create nonpolluting, domestically sourced alternative fuels could put more stable revenue projections in their business plans and be able to raise money more easily. Ultimately they would be able to go to market more easily.

Think of what you heard in the Senate hearings about bailing out Detroit. One argument was that Detroit "wasn't ready" for the shift to small cars in 2008. They weren't, because no energy policy like the Green Dividend had been in place to allow auto executives to count on rising fuel prices. Like their Japanese competitors, they should have been ready for this shift even without such a policy, but these auto executives are not rocket scientists. (If they were, Lake Okeechobee and the Mojave Desert would be full of fuselages.) It's not just auto makers, drivers and renewable-energy firms that would appreciate greater certainty. Consider other benefits: Municipalities, trying to decide whether the current interest in mass transit is likely to be sustained now once people get used to paying $3-$4 for a gallon of gasoline, would be able to expand their networks without worrying whether current higher levels of ridership and revenues will suddenly evaporate, leaving them with a fleet of underused buses and a higher level of debt service. And all this high-speed rail service you hear about in the Stimulus Package, designed to facilitate travel among our cities (or at least, between Tampa and Orlando)? Such projects are much easier to finance if people expect higher gasoline prices.

If homeowners and developers knew that fuel was going to be an increasing expense in the future, houses would become more efficient and would perhaps be located in more sensible clusters.

In total, the business environment and the purchasing environment would soon become much more user-friendly because the guesswork would be removed from prediction. An environment of certainty would create lower prices for and a greater supply of energy-sparing goods and services, and would expedite the adoption of conservation measures.

3. *It is easier to compete against a non-subsidized alternative than against an alternative whose market price includes hidden subsidies.*

While partisan politics is not part of *Why The Heck* (haha, good one, Al), management of subsidies is a situation in which both parties are guilty of economic malpractice. The Bush administration subsidized biofuels and, as noted, its biofuels program should now appear in a thesaurus as a synonym for "unintended consequences." The Obama administration is also heavily subsidizing alternative fuels, albeit more environmentally benign next-generation alternative fuels. Will there be an unintended consequence? Who knows? By definition an unintended consequence is unknown. My suspicion is that we are looking at many more years of subsidies than anyone anticipates. As one of the commenters on *Why The Heck* noted, conventional fuel costs per unit of output produced tend to come down over time (adjusted for inflation), making their price a moving target against possible alternatives.

Is there an argument in favor of continued pricing uncertainty?

Consider the component of the Green Dividend proposal related to maintaining price stability. Even those who oppose a sizeable fuel tax must believe that reducing variability is a good thing. Otherwise, they end up arguing that uncertainty is a good thing. *Au Contraire*. Uncertainty is a cost that people pay to guard against, not a benefit that people pay extra to attain. Nobody except a few speculators prefers a non-predictable future price of fuel. Taking uncertainty out of the equation removes a major "cost" with no trade-off elsewhere. There is simply no benefit to uncertainty other than to people who arbitrage it.

All three of those arguments would lead one to the same conclusions:

- Gasoline prices should be high enough—capturing all of their costs of production including those supply-line defense costs now financed by the Government—that alternative energy and new conservation technologies and trains can compete on a level playing field without more than temporary subsidies;

- Gasoline prices should be stable enough so that companies sponsoring those technologies can take fossil fuel pricing risk out of their business plans, allowing them to raise much more money more inexpensively.

The counter-argument? It's been repeated many times. We can't afford a gasoline tax. (Note that we are focusing on gasoline and not fossil fuels generally because to believe in taxing fuels other than oil, fuels which generally do not come from unfriendly countries, you must believe in global warming. *Why The Heck* is global warming-agnostic. Ask me privately and I will say, yes, I believe in global warming and would tax all fossil fuels. But I won't say that publicly. What happens in *Why The Heck*, stays in *Why The Heck*.)

Another way to look at it: Gasoline is expensive whether the expense is captured in its price or not. All we are debating here is whether the expense should be borne by people who use the product, or by the rest of us. We are already paying a gasoline tax that we can't afford — it just isn't being collected at the pump.

It would not bankrupt the economy to collect it at the pump. How can I be so certain? Recall that we paid a $1-$2/gallon "tax" throughout much of 2008 except that the money was going to the Middle East rather than to our own government, and therefore was not offset by reduced taxes elsewhere, and guess what? The economy didn't figuratively "go bankrupt" until the "tax" came back down.

Even so, remember that we aren't trying to justify imposing a gasoline tax. There is *already* a tax, whether we pay it at the pump or not. Raising the price to cover the costs is economics. Returning that extra cost to people explicitly to make the whole thing revenue-neutral is *Why The Heck*. That's why instead of a conventional gas tax, we are trying to justify a kinder, gentler version of it — a "Green Dividend." A Green Dividend differs from a conventional gasoline or fossil fuels tax proposal in four key ways:

1. The money raised from the higher gas tax is clearly paid out to all of us in advance, so that people can easily see the savings offsetting the gas tax. Let us coin a new term for this concept of a rebate paid in advance — a "Prebate."

2. Even those tax proposals which include tax offsets normally focus on *barely visible* tax offsets, like Social Security tax reductions, that are hard to see. That type of proposal pits a highly visible tax increase against a subtle tax offset. Not a recipe for widespread public acceptance.

3. The "branding" makes a Green Dividend even more visible and the need for it even more apparent.

4. The tax varies inversely with the price of crude oil so that the gasoline price at the pump stays stable, plus a built-in predictable increase over time, as we try to wean ourselves off of it. And if the price of crude oil were to spike on its own again, something that rarely happens after a bubble bursts, the tax could even be lower than it is now or could be temporarily waived altogether, just as two presidential candidates were proposing during the summer of 2008.

That last paragraph is worth repeating in slightly different words. If you remember only one thing from this entire book, this is the thing to remember: *A highly visible tax must be offset with a highly visible "prebate," a rebate in advance.* Just saying, "Yeah, we'll reduce other taxes to offset it," won't work.

Green Dividend vs. High CAFE standard: Which more effectively reduces fuel usage?

The Corporate Average Fuel Economy (CAFE) standard is going to rise again, to 39 miles-per-gallon by 2016. Does such a high CAFE standard negate the need for a Green Dividend by making vehicles more efficient? (By way of background, a CAFE standard penalizes carmakers whose fleets are fuel-inefficient on average. That penalty presumably gets reflected in higher prices for models which get poor gas mileage.)

Far be it from anyone here at *Why The Heck* World Headquarters (meaning me) to argue with anything that should discourage gasoline use. Having said that, a CAFE standard, like any other non-market mechanism designed to substitute for a market signal like a price change, creates distortions and/or doesn't quite accomplish its goal. In this case, the CAFE standard doesn't apply retroactively and higher fuel efficiency doesn't discourage driving, once the car or truck is purchased.

Paradoxically, a stiffer CAFE standard could even encourage driving (and increase traffic) once the car is purchased, because better fuel efficiency reduces the variable cost of driving per mile unless the higher fuel efficiency is offset by higher fuel prices. A lower cost of driving means that mass transit and Amtrak will continue to be the unwanted stepchildren in the transportation family. The government can speed trains up all it wants but as long as driving is cheaper and is fast enough point-to-point, it won't get people out of cars.

The tax on large new vehicles will also encourage people to keep large old vehicles on the road longer. In 2009 the government tried addressing that problem with *yet another* large subsidy, paying people to junk their big old vehicles if they replaced them with vehicles getting at least four (!) miles-per-gallon more, and people did...but the vehicles they bought were mostly made overseas.

Because CAFÉ standards do not change the price of gasoline, the opportunity to create a new sector in alternatives to fossil fuels will be lost or at least compromised.

Finally, the size of the CAFE tax on inefficient new vehicles is nowhere near proportionate to the size of the subsidy on gasoline, so that even with a penalty or tax paid for not meeting mileage standards, far more large vehicles will be sold than is economically optimal.

There is no reason for selecting a CAFE-based intervention over a price-based intervention other than that the Obama Administration considers a gasoline tax to be politically unpalatable—mostly because they haven't considered the Green Dividend option to make it palatable. There is also no reason that the latter and former should be considered substitutes rather than complements.

And that's it. A high, stable, and gently rising gasoline price offset by annual, extremely visible, environmentally friendly Green Dividend checks. Now let's work out an arithmetic example and then some details.

First, an example: Suppose the average driver uses about 1000 gallons of gasoline per year, and the starting gasoline tax was set at $1.00 a gallon, so that the tax would raise the average person's cost of fuel by $1000. By sending every driver a $2000 check to offset this $1000 tax, any driver using less than 2000 gallons (about 35,000 miles/year's worth or more) would be "made whole" even if he or she purchased the same amount of gasoline with the tax as before the tax. Common sense, basic economics and experience strongly suggest that most people would try to make do with less fuel as the price rose.

Each year the size and possibly the scope (in terms of covered fuels) would increase, offset by a higher check. However, because driving would decline, as it did when gasoline prices spiked in the summer of 2008, gas tax revenues would not rise as fast as the gas tax itself. That means the Green Dividend check would not rise as fast as the tax per gallon of gas, and that's actually a good thing. People would be driving less, getting better mileage and spending less on fuel.

Especially because we are still in a Methadone Economy and need our annual stimulus fix, "revenue neutrality" is not a central requirement of stimulation policies. Hence the size of the check could be large enough so that most people—even if they drive the same amount in the same car despite much higher gasoline prices—would be no worse off than today. Most people would drive less or switch to mass transit or a smaller car, and would be better off. Of course, there will always be a few people who are worse off with any policy change, but most people were made worse off by $4/gallon gasoline not offset with any dividend of any color but survived. Green, as luck would have it, is the color of Islam, so in 2008 it was largely the Islamic countries of the Middle East that received a Green Dividend.

Obviously the greater the number of people whose fuel expenses are offset or more than offset by this check, the greater the net expense to the Government, but whatever the size of the Green Dividend, it would pale in comparison with the long-term cost of doing nothing.

2011 might be a perfect year to do this, as the extension of the Bush tax cuts are debated.   A compromise position between liberals who want to reduce fossil fuel usage and conservatives who don't want to raise taxes might be to restore the income taxes themselves, but offset the increases with a Green Dividend.

Doing nothing means more dependence on imported fossil fuels from unfriendly or unstable countries, more money allocated to protect against them, more traffic and pollution, more cost for auto companies trying to guess which cars to make, more fluctuating prices and inability to project future trends, and more government subsidies at taxpayer expense for alternative fuels which will not be cost-effective anytime soon.

This is not speculation. This indeed has been our *de facto* energy policy since the early 1980s.

Why high and steady fuel prices are exactly what the country needs to get out of the recession...and stay out.

Now let's talk Big Picture and explain my earlier comment that "the sustainability of the recovery depends on fuel prices. They need to go higher and stay higher."

The country needs a new industrial sector, not just a few new companies, to begin the next boom. Removing all hidden and indirect subsidies from gasoline and eventually other fossil fuels, and instead keeping the price "where it should be" to cover the full cost of gasoline yields a level playing field for investments in alternative fuels and conservation. That level playing field would, with only temporary subsidies to get started, facilitate the development of the next generation of technologies. If it happens in this country, this technology development will put the U.S. at the cutting edge of the science and create many knowledge-based jobs. It isn't going to happen if we continue to allow fossil fuel prices to fluctuate at low prices.

Pay attention here. Look at the economic history of the United States. Each sustained boom in our history was led by major sectors. Not just a few industries, but entire sectors. Agriculture, transportation, manufacturing, technology, the Internet. It's a good bet that the next sustained boom isn't just going to happen because enough economic Methadone causes people to start spending again on largely imported consumer goods.

Ah, yes, consumer spending, the barometer everyone looks at to track economic growth. Indeed, increased consumer spending did generate economic growth during the first decade of this century following the dot-com bust. People did start spending again based on feeling richer because of rising home values, but *without a major new industrial sector to support true economic growth, the spending increase could not be sustained*. We should have learned that, as in past sustained expansions, only true wealth creation can drive sustained spending increases, and only new sectors drive true wealth creation. The best bet for such a sector today could very well be renewable energy/conservation.

Wealth creation won't happen if this sector has to rely on government largesse, doled out to the idea *du jour* based on which firms have the best lobbyists, because the renewable/conservation sector is competing against heavily subsidized fossil fuels. Investors won't want to bet on businesses whose economics depend on Congressional largesse. Instead, level the playing field and watch what happens to this new sector.

A pipe dream? No, a best practice. One need only look at Denmark to see what happens to a country that raises the price of gasoline to roughly...hmm...in this case, I'm not sure any of you are ready for this figure, so I will refer you instead to another book, *Hot, Flat and Crowded* by Thomas Friedman, which tells the Denmark story in detail. Suffice it to say that Denmark developed an entire renewable energy sector through high domestic fuel taxes, in lieu of higher taxation on most other things. You know those windmills you see occasionally dotting our landscape? Well, which country do you think produced them? Is it because the Danes are smarter than we are? Heck, no. It's because they consciously decided to be the first kids on the block to develop this new sector. Unlike the U.S., where the energy policy is designed to subsidize fossil fuels at the expense of growing a renewable energy/ conservation sector, the Danes, with an eye towards the future and no domestic oil industry to lobby against their policy preferences, are doing precisely the opposite.

If each country were a stock and you were an investor, which would you buy? Oh, and they developed this sector without any significant home market to speak of to support scale economies. The population of Denmark is about the same as that of Wisconsin, only thinner. The U.S. has the world's largest domestic market, which should make developing this sector easier for us than it was for them...provided we level the energy playing field first.

While in the short run people will buy less fossil fuel because the price will increase relative to everything else, ultimately they will buy less fossil fuel because—thanks to an environment of certainty creating the ability to plan and invest accordingly— they will *need* less fossil fuel. This is the outcome we all want to achieve, and this is what the Green Dividend promises, perhaps not totally painlessly but much more painlessly than any other way of getting from here to there.

Whatever pain there is, except to outliers who drive a great deal, will be caused by the inevitable imperfections presented by the details. How high should the tax be and how quickly should it cause the underlying price of fuel to rise (offset by higher green dividend checks each year)? When, if at all, does the Green Dividend shift from covering just gasoline to all carbon-based fuels? How is it determined who gets the checks—all drivers or all taxpayers or all adults? What about commercial users? How often are the checks sent out? And how are existing differences in gasoline consumption by state taken into account?

Yes, these are real issues and there are many more of them, no doubt. But these political and logistical issues should not suffocate a debate over the Green Dividend, just influence details of its outcome. They are secondary effects that will never be addressed perfectly because no government intervention in a marketplace, however well-targeted and well-designed, gets an "A" for equity or efficiency. This is true no matter what mechanism one uses to influence fossil fuel prices. For instance, current energy policy favors high utilizers over low utilizers and therefore is also inequitable and inefficient.

Nonetheless the potential for backlash over the details should not obscure the big picture that the Green Dividend turns a remarkably inefficient and counterproductive subsidy and an unappreciated but almost equally counterproductive "cost" of uncertainty into a source of economic advantage so significant that it could by itself figuratively and literally fuel the next sustained recovery.

COMMENTS

This posting ranked second in the number of comments, after the posting proposing blowing up the real estate brokerage cartel. (It probably ranked first in comments by non-real estate brokers.) I've categorized the comments as "positive" or "negative" but it would be more accurate to term most of them simply "helpful."

POSITIVE COMMENTS

The first positive comment came from someone calling himself WhytheHeckFan (and God bless him) observing that these checks would present a perfect opportunity to attach a ShopAmerica Gift Card, making the Green Dividend even more palatable.

Another comment came from someone who had the previous summer paid a premium over sticker to buy a hybrid on the assumption that gasoline was someday going to cost $5 a gallon. This is exactly the point about the importance of keeping fuel prices predictable. He is stuck now with high car payments and lower-than-expected fuel cost savings, his reward for trying to do the right thing in a country that he says "hasn't had an energy policy since the Carter Administration."

Two comments pointed out variations on the same theme: that 90% of the impact could be accomplished simply by *telling* people that this proposal is a certainty to happen in, for instance, 2012, and giving them a few years to adjust.

Someone else wondered where Kleiner Perkins is on this. Arguably the best-known venture capital firm, they've made a big bet on alternative fuels but, thanks to the absence of a variable tax holding the price of carbon-based fuels at the level as when they made these investments, most of those investments are pretty much underwater now.

Someone else observed that if President Bush had announced in Summer 2008 "We'll suspend the federal gas tax now but if the price of gasoline ever falls below $3 we will put enough of a tax on it to make that the permanent floor," virtually no one would have objected. However, that would have required Mr. Bush to undertake something called "leadership," which was right up there with "eloquence" on the list of attributes not often associated with him.

Yet another suggestion proposed intensifying the Green Dividend effect with a vehicle tax tied to EPA mileage estimates or weight. And indeed a "mileage tax" is being piloted in Oregon.

One graduate student in economics introduced me to the concept of the eponymously named "Hotelling Point," the point at which alternative energy becomes competitive with traditional sources. His comment on the blog:

> The whole alternative energy industry is in danger here
> because the Hotelling Point never seems to get any closer.
> The price of regular energy will fall over time, like it has
> ever since the Greeks lit their lamps with olive oil. And

crude oil costs less today than it did in 1980 even before adjusting for inflation. The answer in Washington seems to be to subsidize alternative energy rather than make regular energy cost more, and try to chase the Hotelling Point downwards. That is expensive and inefficient because it will discourage conservation no matter what source of energy is used.

A generally positive comment also noted that one of the pillars of *Why The Heck* is that compliance is "voluntary," and there doesn't seem to be anything voluntary about a gas tax. I beg to differ. Recall that in addition to paying a tax, you are also receiving a "Prebate" of equivalent or greater value. Deciding whether to drive so much that you will lose money on this transaction is indeed voluntary. If it were just a tax alone, then, yes, it would not be voluntary. The Green Dividend offers a choice. Most people would save money by simply driving less or driving a smaller car.

Someone else noted that consumer spending would be boosted if people used their Green Dividends to invest in energy savings—especially if and when such a tax is extended to all fossil fuels. At the risk of sounding judgmental, there is good consumer spending and bad consumer spending. Investments that reduce future energy costs would be "good" consumer spending, while supersizing your Big Mac might not be.

Perhaps—this is my own comment—the ShopAmerica Gift Cards that go out with the Green Dividends could be limited to energy-saving devices.

One commenter wrote: "The Green Dividend is to a gas tax what Flintstones vitamins are to cod liver oil." True, and you only have to look at "cap-and-trade," which is like a gas tax on utilities burning coal, to see that. One salient feature of cap-

and-trade is that coal-burning power plants would have to purchase the right to pollute from more efficient plants, thus raising the price of most electricity generated from coal. Cap-and-trade as it stands now won't boost consumer spending on conservation remotely as much as a Green Dividend would, because it won't be accompanied by an explicit check. Fact is, government revenues raised from cap-and-trade are taxes that don't have to be raised from other sources, but the lack of an explicit tie-in with a Green Dividend will make cap-and-trade less popular than it should be, at least to those who believe that global warming is real.

By the way, the other reasons we aren't emphasizing cap-and-trade are (1) we didn't invent it — all the ideas in *Why The Heck* are at least somewhat original and (2) most of what's burned for fuel by utilities is produced domestically, so the best argument for cap-and-trade is to reduce carbon dioxide emissions, and the Green Dividend is about reducing dependence on foreign oil, not about global warming.

Next came the observation that the difference between this boom and the last real wealth-creating boom in the dot-com sector is that this boom, though likely to generate more new high-paying jobs in high-end engineering, manufacturing, and construction, won't increase productivity of the buyer remotely as much as the Internet did. Energy is a commodity while the Internet represented almost a new way of life. People and businesses will save a little money on energy as new forms of energy become less expensive than fossil fuels, but slightly cheaper energy won't revolutionize the way people conduct their lives.

Two final observations along one theme: Many mainstream commentators have said "we need more green energy" but none of them has followed that with an observation that dirty fuel is too cheap for green energy to be competitive. The mainstream answer is always for the Government to subsidize the green fuel, never to tax the dirty fuel. I am getting really tired of solutions that begin with the phrase, "the Government should subsidize." They show a lack of creativity that really makes me want to puke. That's why subsidies are never, and will never be, part of *Why The Heck*.

CHAPTER TEN

Reducing the Cost of Being
Poor Even More

The government should subsidize bank accounts for poor people.

Yeah, yeah. So I voted against subsidies before I voted for them. Call me John Kerry. Sic the Swift-Boat Veterans on me to prove that I never set foot in Vietnam, that I once lied about my age on Match.com, and that most of the chapters in this book were in reality written by Sir Francis Bacon.[29] Yes, I flip-flopped. It's also called "changing one's mind after receiving new information or developing new ideas" and if nobody ever did that we'd still be ruled by the British.

Really it's more like "making an exception" because *Why The Heck* philosophy generally remains anti-subsidy. Here,

---

29. Normally I wouldn't insult the intelligence of my readers by explaining cultural allusions but this one is pretty obscure even for *Why The Heck*. Bacon is the guy who some people think wrote Shakespeare's plays.

however, *Why The Heck* proposes one instance in which a government subsidy makes a lot of sense—and we will also propose how it can be very profitable for the Government too. A bank account is not just a convenience. It's one of those things, like your voice, that you only learn to appreciate when you lose it. Take direct deposit. You don't think twice about it. Your paycheck goes to your bank account and then you draw on it.

But what if you don't have a bank account? You can't have direct deposit. You go to a "checks cashed" place. Think those are free? Think again. As part of doing actual research, I went to one to cash a check. They wanted to give me ninety cents on the dollar. That's 10% to get my money now instead of waiting five days for the check to clear. The annual percentage rate I would be paying to get my money now would in effect be four figures. All because I don't have a bank account.

There must be a way to arbitrage between a four-figure interest rate, or other high rates as described below, and a bank account. Suppose the Government gives everyone who wants one a $300 bank account, but charges back $3 for the first 150 transactions. Everyone benefits except the check-cashers who would have charged perhaps $10 for such a transaction, but the Government is still out $300 per person. When multiplied by millions of people, that is a lot of money no matter how worthy the cause.

But remember that the balance cannot fall below $300 until 150 deposits have been made and/or 150 checks have been cashed. Over time, the Government would get $300 out, plus $150 more, by charging the account holder $3 for each transaction.

$450 back on $300 invested is a healthy return even if it takes a few years.

To make it fair for the banks, which presumably would be allowed to charge only a little for ATM transactions, banks could charge extra for teller transactions.

The $300 is intended to provide a buffer for the banks against bounced checks or other exigencies. (The number $300 is chosen because most checks for the working poor will be under that limit and hence cashable.) Very few people who didn't need a bank account would sign up for the "free" $300 account because not being able to draw down the $300 until one has paid 150 three-dollar fees makes it attractive only to those who don't otherwise have an account.

In particular, the next idea proposed in this section reduces the financial burden of being a member of the working poor far more than any income tax cut and probably more than any Social Security tax cut...but without a bank account, the logistics become almost insurmountable.

If there were bank accounts or some ersatz version thereof for everyone, the Government could stimulate the economy and put more money into the hands of both the working poor and small businesses, while reducing both the foreclosure rate and the federal deficit, all without raising taxes.

Hey, don't forget to propose some ideas for us middle class types. We need money so we can afford to buy more copies of your book.

Well, since you put it that way, how can I refuse?

If responsible use of credit is a football, then the credit card industry is Lucy Van Pelt and the American consumer is Charlie Brown. So it's no surprise that the average American adult owes almost $5000 in revolving debt. At the current 20% interest rate, over five years that debt costs about the same amount in interest as the actual principal you owe. That means at the end of five years, $5000 in payments later, you still owe the same $5000 in principal.

The *Why The Heck* proposal: Create a one-time opportunity to borrow from your own pension plan free of charge to pay off your credit card debt, and then repay your own pension fund over those five years. At the end of five years—making about the same monthly payments you would have made to the credit card company—you will have paid off the entire debt. (If you can't repay your pension fund all the money you borrowed, the shortfall becomes early-withdrawal income to you...but even with the early withdrawal income tax penalty, you are probably still ahead of the game.)

At *Why The Heck* World Headquarters (the Barcolounger in the corner of my living room), the entire management team (me) is quite adamant that this opportunity be allowed only once, though. We've all made mistakes by living beyond our means, but just like when Ward would call Beaver into his den to make sure that The Beave learned his lesson ("Gosh, Dad, I guess I learned that it's never OK to borrow money from Mom's purse, even if I'm sure I'm going to return it later"), we will not make the same mistake again.

We don't want to become a nation of Eddie Haskells.

Economic alchemy like the Stimulus Stamps, you say? Read on.

Chances are, if you in the demographic that is reading *Why The Heck*, you have never applied for, and probably never even heard of, the "payday loan." A "payday loan" requires the borrower to assign his next paycheck to the lender in exchange for receiving about 85% of that paycheck in cash immediately. A payday loan is an expensive but dependable lifeline for millions of people[30] who live paycheck to paycheck, perpetually playing catch-up on mortgage and credit card payments.

An 85% payout on a two-week loan works out to such a high annual percentage rate (APR) that in California, for example, the allowable ceiling is 459%. A typical $45 charge on a $300 loan is 360%. The reason that the APR is so high is not just that some people have no other choice and don't understand APRs. It's also that the cost of running these loan stores, doing these microloans one at a time, person-to-person, is high enough that, even with usurious APRs, the lenders rarely live in luxury.

Congress recently considered the "Payday Loan Reform Act," which would have put price ceilings on these loans. The effect of that, of course, would have been to drive borrowers underground. It would have made terms even more onerous because, as we learned during Prohibition, you can't legislate markets out of existence. This bill embodied the precise opposite of *Why The Heck* philosophy.

Payday loans illustrate the differences among Republicans, Democrats and *Why The Heck*.

Republicans generally opposed the Payday Loan Reform Act, preferring working people to get ripped off than to interfere with the functioning of what to them appears to be a perfectly functional market. The majority of Democrats supported the legislation, even though it have dramatically reduced the number of these stores and made legal payday loans almost impossible to obtain in some areas.

---

30. By now you are probably expecting no actual helpful information in these footnotes and indeed one would be excused for wondering why you are continuing to read them at all, but finally your patience will be rewarded with an actual piece of data: Roughly 12-million people close about 70-million of these loans in a typical year.

Why the Heck would look at the market through a wholly different lens. Clearly, a 360% APR is a problem. So far, that viewpoint aligns with the Democrats. But a brief look at the lenders' demographics suggests that this problem is not the fault of greedy, monopolistic lenders. Instead this problem results from the way the industry historically developed. Those of you who remember *A Beautiful Mind* recall that there can be two or more equilibriums in a marketplace, two ways in which a market can function and be stable. In this case, the industry grew up in a pre-electronic age where the optimal equilibrium for this business like most others was a small, staffed storefront—a quaintly retro model in an era when most people with means borrow electronically, on a credit card or home equity line of credit.

If there were some way of reaching a new equilibrium that incorporates automation and operates on a much larger scale, it would be possible to dramatically reduce the price of these loans without reducing availability. Surely there is an arbitrage opportunity between the Government's very low cost of capital and 360%, enough so that the employer could have a stake in the outcome as well. An employer whose employees had access to low-cost payday loans would attract and retain more staff, other things equal. (Generally today employers don't get involved at all in these transactions because of their seedy nature.)

So let's throw the three inputs of technology, low cost of capital, and employer involvement together into the *Why The Heck* cauldron to see what bubbles up.

Imagine a solution where the Government charters (and receives a share in) a number of corporations (enough to compete vigorously) that provide what might be called "ePayday loans," to distinguish them from the old-fashioned corner store payday loans, borrowing at the Government's cost of capital. As part of receiving their charter and low-cost capital, these corporations commit to lend at, for instance, 95 cents on the dollar instead of 85 cents. The APR at 95 cents is still high, but nowhere near 360%.

Now let's bring the employers in. These new corporations would have the electronic capability of contracting directly with employers, facilitating the paycheck advance on a large scale and automating it, direct-depositing the loan into the employee bank account. Sure, this could be done without bank accounts, using a Western Union or other office, but much of the efficiency would be lost and the rates wouldn't be as attractive. Hence the earlier suggestion about the bank account subsidy.

The employers put up some security in order to participate, to cover any defaults by employees. In exchange, they receive some of the profits from the loans. These profits encourage employers to offer inexpensive payday loans as a "benefit" to their employees, making their employees aware that they can now get advances for 95% of their paychecks. (This 95% will fluctuate according to market conditions, of course — it is for illustration only.)

Incentives are now aligned. The lending corporations can direct-deposit loans into borrowers' accounts. Default rates are reduced dramatically because employers put up security. Employers get perhaps 1% rebates on each loan, which encourages them to steer their employees away from streetcorner moneylenders towards the automated direct deposit of loans. Most important, employees can then access these loans, at 95 cents on the dollar rather than 85 cents, from any ATM.

The income of the working poor who regularly use this financing mechanism immediately rises 12% — an amount which for some will exceed the benefit of any conceivable tax cut, but at no cost to the Government. It's a good bet that the entire 12%[31] "windfall" gets recycled into the economy within the week, thus stimulating the economy. Some of this money would be used to pay mortgages (or to pay rent to landlords who themselves have mortgages to finance), thus reducing the foreclosure rate. A good guess is that most of what people receive through a payday loan is paid out right away. Otherwise, why take the "haircut" to get the loan?

Finally, these chartered corporations, partly owned by the Government, should be quite profitable themselves, as they would have access to low-cost federal capital. Negative surprises would be unlikely because lending practices would be carefully monitored to prevent any Fannie Mae type debacles. (Such debacles would be hard to imagine in any case because the loans would be outstanding for only two weeks.) Once the market has reached its new equilibrium, *Why The Heck* philosophy would suggest that the Government sell its stake and exit the market altogether, which should net a nice profit for the Government and therefore cut a bit out of the deficit.

Who loses? The streetcorner moneylenders. And, unlike you, they are not in the demographic likely to read *Why The Heck* so I won't have to change sleeping locations every night.

COMMENTS

---

31. Having spent the whole introduction to Part Two whining about consultants who can't do math, I feel the need to point out why moving from an 85% payout to a 95% payout represents 12% rather than 10%. It's 95%/85%, not 95% - 85%, that determines the percentage increase.

This chapter proposes two linked initiatives. Most of the comments concerned the ePayday loans, so I lumped all the checking account comments together at the end of the section.

POSITIVE COMMENTS

One commenter asked why large corporations couldn't provide e-payday loans themselves, thus eliminating the middleman as well as this chapter. According to John Hoffmire, who as mentioned runs the Center on Business and Poverty (www. cobap.org), some of the large corporations signed up with COBAP do indeed already provide these loans, and there is no reason others couldn't as well if they had access to inexpensive capital, which many do not right now.

Another comment suggested that the employers that sign up to participate could assign their employees' vacation and sick leave time as collateral to the ePayday loan corporations as protection against default. A pleasant thought indeed, if only people in many of these jobs had sick leave or vacation time accruals. They are lucky if their employer gives them workers compensation, which of course is required by law.

Still another commenter observed that these loans are so usurious that some states have legislated them out of existence, which only serves to drive them underground. Those of you with good memories should remember an *Why The Heck* cornerstone: One cannot legislate markets out of existence, particularly markets born of pure desperation.

NEGATIVE COMMENTS

Someone challenged the basis for the ePayday concept being the government's cost of capital. He pointed out that any market would have more favorable economics if the companies in it could borrow at the Government's cost of capital. That is true. The reason that the Government has a lower cost of capital than corporations is that the Government can borrow at the "risk-free rate" whereas corporate debt carries risk for the lender, and hence carries a higher interest rate. The higher interest rate raises costs for the borrower and hence prices. Perhaps the ePayday loan proposal should be that the newly chartered corporations would be able to borrow at the Government's rate for only a certain period of time to establish a track record, and then, once the Government sells its stake, the corporations raise money on their own. Access to government capital is not critical to ePayday loan economics in any case. The majority of the economic advantage is based on automation, scale and employer participation.

COMMENTS ON CHECKING ACCOUNTS

The checking account proposal attracted several comments too. One was from someone who supports his mother. She doesn't have a checking account and lives in another city. In that situation the whole notion that we take for granted of "I'll send you a check" becomes a total hassle. He ends up sending her funds via Western Union, which itself charges roughly 5% per transaction. He said that this posting spurred him to open a checking account for his mother in an amount greater than the check he sends her, but told her not to draw on it, so that he could send her checks which the bank would cash because the bank account balance would cover any shortfall. On a one-on-one basis, he did exactly what the first part of this chapter proposes.

Next, someone pointed out that there is also a public education component. Many people eligible for Medicaid don't realize they are eligible, so it is likewise possible that many people who would be eligible for this program would not realize they had this subsidized checking account option. That is no doubt the case, and there would be some public education requirFinally, someone questioned if $300 would be enough to bring someone into "the system." That is a fair question and I don't know the exact number needed. Perhaps the program could be piloted in different cities with different numbers.

Why the Heck Aren't we Already Doing this Stuff

CHAPTER ELEVEN

Send Us the Richest Refuse of Your Teeming Shores

Why The Heck is not a book about Keynesian economics. However, that doesn't mean that the field of *Why The Heck* repeals the laws of supply and demand on which Keynesian macroeconomics is based. Nosirree. Those laws remain are still strictly enforced.

Here is a brief overview of macroeconomics and a description of why macroeconomics-based deficit spending is the economic equivalent of Methadone, and is not adequate to the task of reinvigorating the economy. Demand for goods — and Keynesian economics is all about demand, rather than supply — comes from four places (to save wear and tear on your eyeballs, we will conveniently abbreviate these sources of demand with the same letters real economists use, in parentheses):

1. Consumers ("C") such as you and me who buy things to keep the economy going, though I must admit I haven't

been pulling my weight in this department for a few years and face it, neither have you

2. Business investment ("I")

3. Government spending ("G")

4. Total exports ("X") minus total imports ("M"). The difference, called net exports, is abbreviated "X-M." Like George Steinbrenner's approval rating in Boston, this difference has been a decidedly negative number for almost 40 years, with about the same probability of turning positive during this geological epoch. (I originally wrote this before he died, but my prediction still holds. Up here, even the obituaries were nasty.)

For an economy to grow, these numbers on balance have to grow (except Steinbrenner's approval rating). Ours is called a consumer-driven economy, which means that rising population and income and, until recently, willingness of consumers to carry increasing amounts of debt, have always made "C" the most reliable source of peacetime economic growth. This, of course, assumes that the consumers themselves have rising incomes over time based on true wealth creation in the economy.

"C" is the most fun way to grow the economy, too, because by definition it involves buying stuff to consume, such as Lindt Excellence Intense Orange Ultra Thin Dark Chocolate Squares™. This particular product is painstakingly prepared with a unique and exclusive blend of only the finest, freshest, purest, highest-quality nouns and adjectives. I have first-hand familiarity with these things, because in sharp contrast to my recent admission that I have been shirking my patriotic duty to consume, I just devoured an entire 6.35-ounce box of 'em <burp>. I didn't even notice I had done so until they were all gone, no doubt because they are so Ultra Thin and contain such large quantities of Excellence.

"I" doesn't taste as good as "C", but it is considered a better source of growth, because it is about The Future. We have made the argument that "I" is critical to wealth creation and the economy won't recover without it. Still, "I" requires a level of confidence in that very same Future that we have not regained yet, so you can't count on "I" for much near-term spending stimulus – it will bounce around for years to come. Also, the critical thing with "I" is to spend it in the right places to create sustainable future growth, which means the Government should not distort the market with subsidies that cause people to overinvest in their own houses, and underinvest in green technology because the price of fossil fuels is too low.

While most people agree that "I" is a good thing and that "C" is too, within reason, there is a lot of controversy about "G." "G" can be good or bad depending on a number of factors, like whether "G" is being spent on education, or on a $27,000 study to determine why inmates want to escape from prison. (Google it.)

Now let's look at net exports, or X-M. The "X" portion of "X-M" isn't really within our control, because it depends on demand from other countries, while "M" for the most part can't be reduced without fiddling with tariffs and quotas. Domestic Content Labeling would help reduce "M" too, as indicated in Chapter Five. For instance, how many of you even noticed that my "C" of the chocolates was really an "M," a Unique Creation of Lindt's Master Chocolatier™...who happens to live in *Switzerland*? So I didn't do a darn thing for the U.S. economy just now, with the exception of Stratham, N.H., where they distribute these Elegant Flavors, Fine Textures and Lingering Tastes, the Ultimate Union of Force and Finesse™.

However, if my box of chocolates had noted the information about its country of origin in larger letters and numbers than it noted the distributor's state of origin, I would not have chosen this Masterfully Balanced Full-Bodied Sophisticated Sensory Experience™. Instead I could have gotten equal value by eating a Hershey bar while reading a thesaurus.

Because "X" and "M" are subject to these vagaries, when "C" and "I" fall off because of the business cycle, Keynes' big insight was that "G" should make up the slack. The Government would stimulate the economy by temporarily running a deficit, spending money to jumpstart the economy. Through the "multiplier effect," money initially spent as "G" gets re-spent on "C" and "I." (Chapter Five, Kinder, Gentler Protectionism, emphasizes the importance to that multiplier of spending that money internally so it doesn't get lost to the economy as "M", like my Ultra Thins just did.)

Normally when "C" and "I" are down, the Government would compensate with more "G," in the form of temporary deficit spending. But what Keynes didn't anticipate — the major fly in the Keynesian ointment, the fallacy in his theory that will prevent him from achieving lasting influence for his economic ideas — was George W. Bush. Mr. Bush transformed the surplus he inherited in 2001 into a deficit by implementing a tax cut, in order to stimulate spending, encourage investment, and redistribute money from future generations of Americans to the current generation of contributors to the Republican Party.

Among the other effects of that deficit, the four most relevant to this discussion are as follows:

- Because "C" was already high even before the tax cut, the resulting increase in "C" drove total consumer spending to unsustainable levels, thus exacerbating the peak-to-trough effect of the 2007-2009 downturn;

- Because the deficit was created by tax cuts and the Iraq War rather than infrastructure investment spending (examples of the latter would be the GI Bill or the Interstate Highway Act or even the space program, which had many civilian spinoffs, including some that didn't taste like oversweetened rehydrated orange pulp), this deficit left no legacy of improved American productivity; and

- The actual numerical deficit grew to a point that limited the ability of the current administration to increase it. Some say the specific incremental deficit spending should have been twice as high as it was, but that would have been economically dangerous because the baseline deficit was so high to begin with.

- Finally, a good deal of that "G" — both in the previous and current administrations — took the form of bailouts, which don't really have a stimulus effect because they just replace previously lost money.

Here's where we stand, then:

- "G" is tapped out. The deficit is simple too high to add more spending.

- "I" suffers from the three blows of excess existing capacity, no new investment spending in any industry except Yankees, and lack of available credit for new companies. And what had been the most reliable source of "I," residential construction, has dried up.

- As for "C," consumers are saving their nickels instead of spending them, paying down their past "C" indulgences, indulgences that once upon a time were believed to count as sustainable spending growth. Incomes and wealth are also down, so for the foreseeable future consumers will be spending a reduced portion of reduced incomes and wealth. Hardly a recipe for growth in consumption.

- And don't get me started on "X-M." I am so sure "X-M" will always be a negative number that if it ever turns positive you can return this book for a full refund.

So...what's left? Where can we get a stimulus that will boost the economy consistently, but that won't further expand the deficit via economic Methadone and enrich our Chinese creditors at the expense of future generations of Americans?

Perhaps part of the solution to our economic downturn — especially the housing crisis — is right at America's doorstep. Let's let more well-to-do immigrants into the country, to buy and rent unused and foreclosed housing units, and to start spending money.

Not just any immigrants, but immigrants with some means. Surely there are fairly well-to-do would-be immigrants out there, working-age people who would pay perhaps $10,000 as an "entry fee" to immigrate following the obvious background investigation, and who can demonstrate a net worth many times that amount (with help from family contributions). In addition to raising money for the Government, the "entry fee" and means test would weed out people who intended to immigrate just to live off public largesse, thereby satisfying those who oppose immigration for that reason.

How many of these people are there? No one knows. But add together the enormous numbers who enter the immigration lottery every year (which itself costs the winners about $1000 in processing fees), plus those who pay middlemen much greater sums to enter illegally, plus those who haven't even tried to enter because the process is so cumbersome and the outcome so uncertain, and the pool probably exceeds 100 million.

Surely some small percentage of this total could meet these financial requirements. Any noticeable number helps.

This new class of relatively well-off immigrants will stimulate the economy. They'll buy or rent houses. They would also furnish those houses and patronize local businesses, creating jobs and boosting local tax rolls.

Of course, even for up to $10,000 apiece, not just anyone should be allowed to immigrate. We could also use this as an opportunity to address some other labor force issues. Surely there are skill sets that are more in demand than others, as in health care. For instance, licensing requirements could be expedited so that regional shortages of nurses and primary care physicians could be alleviated.

While we're on the subject, Thomas Friedman has often proposed that we seek out qualified immigrants to provide the intellectual horsepower to drive the next generation of investment and assist a long-term sustained recovery. I would agree—those people also bring capital. It's just intellectual capital instead of financial capital. Both are important. The Obama administration has significantly loosened restrictions on the class of visa Mr. Friedman is talking about. This "richest refuse" classification is different, however, and is not addressed by the recent policy change.

Both this new policy and the *Why The Heck* idea are based on the same principle: that this is our country. We make the rules. No law says we can't change those rules to make the immigration system look more like a college admissions process, where we select applicants based on qualifications. Our current system of quotas was put in place years ago and is based much more on politics than economics. We no longer have that luxury.

This is not to say that Mr. Friedman's and my proposals should be *in lieu* of the current system, just in addition to it. While letting the Usual Suspects immigrate is a good or a bad thing depending on your viewpoint, this is not about quotas and border walls and amnesty. This is about sifting through the pent-up demand to live in America, in order to find the people who can do us the most good, whether short-term through their means, or long-term through their qualifications.

It is worth emphasizing one more time that neither my proposal nor Mr. Friedman's has anything to do with the usual "immigration" debate, and therefore should not get confused with it. I know there are many people who immediately decided to oppose this plan based on the chapter title, because they hear "immigrants" and think "immigration."

Instead, this proposal is purely about elementary macroeconomics and the one surefire, untapped way to stimulate the economy consistently over several years. We need immigrants because nothing else is expected to really turbocharge the growth rate.

COMMENTS

Comments on this posting were mostly of the "why aren't we already doing this?" variety and the "it must be getting caught up in the immigration debate" conjectures. The fact is that we are already expanding corporate-sponsored visas for people with certain skill sets, but that is not the specific proposal here. The proposal is to do something similar for people with certain economics. Once again, the Government already does already make exceptions to immigration law for immigrants bringing in lots of money, but the financial net worth threshold is at least a magnitude higher than it should be to provide the stimulus proposed here.

POSITIVE COMMENTS

Someone noted that Canada took steps similar to what is being proposed here when it was faced with a housing oversupply in the 1990s. To this day, Canada has the highest in-migration rate of any country (relative to the number of people already there)…and yet Canada's unemployment rate is usually lower than that of the United States.

NEGATIVE COMMENTS

Once again, the negative comments were more interesting than the positive ones. Several of them, despite roughly every other sentence in the posting saying, "This is not about immigration policy," railed against immigration policy. Other comments said that the immigrants would take our jobs, but the whole point of bringing in people with means and with uniquely in-demand skill sets is that they would *create* jobs.

Another comment doubted that there were a million potential immigrants who would fit the bill. Obviously there is no way of knowing how many there are until the policy is tried, but this isn't like electoral votes. You don't need to get a certain number in order to win. Any number is better than none. There are so many empty houses now that builders in some cases are paying people to live in them in order to make a development look more vibrant.

The two most provocative cautionary comments concerned the potential reaction of other countries. One asked if other countries might object to a brain drain and/or a money drain. The answer is, maybe. However, many countries, such as the Philippines (motto: "The Land That Infrastructure Forgot"), have far more skilled citizens than they have opportunities for those citizens. For instance, in Manila (motto: "Honk if you like Honking"), you need two years of college to work as a counter-person at a McDonalds. We're talking about a country that has some of the world's most educated people and some of its best English-speakers, but whose major industries are rice, pineapples, tourism, component assembly, bribery, reproduction and traffic.

The other more subtle reason many other countries wouldn't mind is that most of the world still depends on the U.S. for its ups and downs. Many if not most countries are directly or indirectly invested in our success. If all they have to do to advance our and their own best interest is to allow a few people to immigrate here, they ought to be willing to pitch in.

These countries wouldn't even notice a few people are missing. Remember *The Great Escape*? You may not, but I am a pop culture maven so I remember the plotline and cast of every movie I've ever seen like it was yesterday. So pay attention and you might learn something: In *The Great Escape*, Steve McQueen, Paul Newman, William Holden and Bob Crane dig an entire escape tunnel without Colonel Klink or Sergeant Schultz noticing, by sprinkling the dirt all over Stalag 17. Likewise, if 50 countries each allowed 20,000 people to emigrate, their economies wouldn't reflect more than rounding error. But it would create a big improvement in ours, which in turn would help lead the world out of this downturn.

Finally, someone noted that many countries have restrictions on the amount of foreign exchange that can be expatriated. Such policies would indeed limit the supply of potential well-heeled immigrants but there are more than 200 countries in the world, only 50 of which would have to be willing to part with 20,000 people apiece for us to get our million.

Why the Heck Aren't we Already Doing this Stuff

PART FIVE

One Final Step to Kick the Economic Methadone Habit

The typical family is finally starting to reach between the sofa cushions to find some money to spend. Perhaps consumers are spending again because the paint is peeling off their house, their car can no longer pass inspection due to the steering wheel having fallen off, or their socks have sprouted holes bigger than many nine-digit zip codes. Still, people are spending a lot less on fun stuff because real growth in standards of living is still elusive...and will only take place like it did in the good old days if as a country we innovate like we used to in the old days.

I emphasize innovation because the Methadone Economy must, by definition, be a relatively short-term fix. Consumers once spent more than they should have, largely because the equity growth in their houses and 401k's made them feel richer, and hence willing to save less and take on more debt. We need to actually create tangible sources of wealth in order to drive future spending.

This chapter is about innovation, not specifically about technology. So there isn't going to be any jargon, and I won't even use any big words. I myself am a bit technologically challenged. For instance, when I was a kid I really did think that Gilligan could receive radio signals after the Professor filled his cavity.[32]

Yet few if any official, high-visibility measurements specifically focus on creating a Sobriety Economy. Consider the indicators government worries about, and the indicators investors await. Most are related to consumers, like retail spending, consumer confidence, housing starts, and auto sales. Others are related to traditional business investment, like factory orders or the Purchasing Managers Index. Few regularly reported numbers specifically measure anything approaching innovation...and yet innovation is what drives the economy. This last section proposes exactly that type of measurement.

Measurement is a pretty dull subject to wrap up with, but the fact is what's measured is what's emphasized, so this chapter lays the groundwork for a sustained recovery creating exactly what everyone is hoping for: high-paying, secure jobs, "green" technology, and book sales.

---

32. I did, however, question the premise that he could change stations just by turning his head.

Chapter Twelve

The Body Count:
Why Measurement Matters

The thesis of this chapter draws upon the old (1986) saying: What Matters is What's Measured and What's Measured is What's Maximized. This chapter will prove the thesis that the Government can sustain a recovery partly just by changing what's measured.

Or you can just take our word for it, skip this chapter and go right to Part Six, to learn how, like Charlie, you can win the Golden Ticket. Then, with the extra time you've saved, you can go quaff a brewski.

For the rest of you, let's start by climbing into the Wayback Machine and setting the dial for Vietnam during the 1960s.[33] The United States lost the Vietnam War for a variety of reasons, which for the most part are beyond the scope of this book mostly because they are beyond the scope of this author. One

---

33. Extra points if you remember what cartoon featured the Wayback Machine. (Sherman & Mr. Peabody, on *Rocky & Bullwinkle*)

reason, though, is well within my scope: We were measuring the wrong thing. Many people in authority thought, until it was too late, that we were winning the war because we were killing more of their guys than they were killing of our guys. We focused on the "body count" on each side, duly reported on the evening news every week.

I even recall having a discussion of the body count in sixth grade. No, not of its reliability, relevance as an indicator, or even of its morality. We were given the two numbers in math class, and were instructed to compute a ratio.

Why were we focused on the body count? Because it was measurable and because we liked the way the measurement turned out because we were more effective at killing people than they were. The fact that this indicator was almost totally irrelevant rarely crossed anyone's mind in the Johnson Administration. One could argue that it wasn't just irrelevant. Like that misleading "red light ahead" sign at the top of the hill in Wellesley, it was counterproductive. It led to a focus on killing people rather than on winning hearts and minds, thus creating a culture that led to the My Lai massacre, as well as to possibly apocryphal military spokesperson quotations like "We had to destroy the village in order to save it."

There was no measurement of the number of good things the military did, like the number of innocent people spared or the number of displaced families housed. As a result there was a lot of napalming there and a lot of righteous indignation here, but for the White House, it was largely about counting bodies. (Whatever else you and I might criticize about the Iraq War, the military has learned one lesson from Vietnam: It now measures a wide range of variables to determine whether hearts and minds are being won. So we focus much more on minimizing our casualties and civilian casualties, and much less on maximizing enemy deaths.)

Meanwhile, the Viet Cong and North Vietnamese were concentrating on what was really important — fighting for unification and independence, understanding and integrating into the local culture, and torturing Sylvester Stallone. Maybe their idea of "integrating into the local culture" was more about coercion than warmth-and-fuzziness, but what they weren't doing was counting the number of our guys they killed.

Yes, I am oversimplifying here. I don't claim to know anything about 'Nam. I do, however, claim to know a lot about measurement and its effect on subsequent behaviors and outcomes. That has been, after all, my "day job" in health care — measuring disease management outcomes. It is true that measurement is not by itself a productive activity, the way "investment" or "R&D" are productive activities. Measurement is a passive activity creating nothing tangible, and by definition it is done after the fact. Nonetheless, I've learned one key thing from doing all this measurement: What you measure looking backward influences or even dictates what you do going forward.

In my field of disease management, for example, most consultants don't know how to measure the actual outcomes of disease management: the number of heart attacks and other medical events prevented by education and support phone calls to people with heart disease. Instead they measure the easiest variable to measure: the total number of phone calls made to people with heart disease, with no concern for whether the calls involved the right advice for the patient's needs, whether the patient takes that advice, and whether that advice prevented a heart attack.

Without being able to point to the avoidance of actual heart attacks, they end up compiling lengthy but irrelevant number-intensive reports with all sorts of measurements of phone calls being made, reminiscent of what Captain Renault said to Major Strasser: "Realizing the importance of this case, my men are rounding up twice the usual number of suspects."

And here's another measurement issue that hits close to home. Ever noticed that you can't buy most books from the authors? Like, try going to www.WhyTheHeck.com to buy this book. Instead of simply selling it to you, I link you to Amazon. Why? Because copies I sell myself don't count towards the "official" sales figures used by booksellers to determine what books they stock and how prominently they display them. So after all these chapters about middleman elimination, I myself am forced to add a middleman to the bookselling process, only because that's what's measured.

Bottom line: Whether it's Vietnam, *Casablanca*, health care, or *Why The Heck*, what matters is what's measured, and what's measured is what's maximized.

What matters in our economy

What matters for our economic recovery — what has throughout history kept us, in the immortal words of the great philosopher Frank Sinatra, "King of the hill, top of the heap" — is our continuing ability to innovate. How do we know that? Throughout history we have invented things and started companies to commercialize them and produce them. As the production of these items became more routinized, the locus of production has moved to lower-cost venues, first lower-cost states and then lower-cost countries. Historically, that industrial migration takes generations. More recently, it has taken only years.

Therein lies one of the reasons the U.S. economy is no longer invincible: The flattening of the globe (to use Thomas Friedman's metaphor) has taken place at a magnitudinally faster pace in recent years than in previous decades. If we don't innovate constantly, we stand to lose much more than we've lost historically, as previous innovations move offshore at an accelerating pace. If you look at the industries where we have sustained our edge and not lost much if any ground to foreign competitors, they tend to be industries where major innovation is constant. In entertainment, for example, each new product is an innovation.[34] Likewise, pharmaceuticals and medical technology never stop innovating. (In both the medical and entertainment fields, it helps to have a large domestic market too, but we also have large domestic markets for other, less innovative products like clothing, and that's all moved offshore.)

The Government often tries to prop up industries in which innovation has become relatively less important and routinization more so, attempting to delay or deny the economic reality that other countries—those which are particularly good at providing equal or better-quality products at a lower cost—are gaining market share. When was the last time the U.S. did this to other countries on a major scale? Garment manufacturing, a century ago. That was the last time the U.S. developed a major industry without being first into the market. (We weren't first in cars when making them was a craft but we were first into making them as a business.)

---

34. Even that song by Steam that goes like this: "Na na na na na na na na hey hey good-bye."

Indeed one would have a hard time pointing to anything other than innovation, in all its forms—including innovation in educational, legal and financing mechanisms to support innovative entrepreneurs—as having been the key to the success of the U.S. economy in our own lifetimes.

To focus on innovation to sustain a recovery, *Why The Heck* has already suggested one idea, which is to stop subsidizing the one major sector, fossil fuels, whose substitutes we are attempting to encourage via innovation. But more broadly, we can't just look product-by-product and decide which ones should be encouraged via subsidies and incentives. Micromanaging innovation is not something governments are particularly good at.

Renewable energy/conservation should be a good bet, but in general we can't assume we know where we need to innovate, or what sectors to encourage. Almost by the definition of innovation, we don't know which technologies today will fuel the next decade's boom, any more than we knew in 1989 that the 1990s boom would be driven by the Internet. But we do know that innovation, in whatever form, will be key to the next sustained Methadone-free recovery just like it has been key to past ones.

If it's that important, why don't we measure it?

Measuring what matters

Encouragement of innovation isn't hard, as compared to other government initiatives. Innovation is what the private sector does, and usually the best government policy is to leave innovation alone. Of course "leaving innovation alone" does not mean "passively getting in the way of it." The Green

Dividend chapter describes how hidden fossil fuel subsidies passively prevent innovation in alternatives. *Why The Heck* philosophy would say, leave it alone...but only after creating a level playing field.

Government can unbalance the playing field in the other direction too, with subsidies. Recall in the Introduction, what has happened to this economy as a result of hidden and explicit subsidies to encourage certain activities, like buying houses.

Ah, yes, housing. Housing exemplifies what can happen when you combine two *Why The Heck* no-nos: (1) subsidies and (2) the absence of effective outcomes measurement. The housing bubble was in fact measured by a prominent Yale economist named Robert Shiller, who tracked the ratio of housing prices to income (and also building costs) going back more than a century.[35] He found that this ratio had been remarkably consistent until this century, when it broke through its previous high-water marks and continued skyward. Surely if that index had gotten remotely as much notice when it was rising as it did when it was declining — if it had been tracked and reported monthly for decades like most other indices commonly used today and therefore the bubble had been much more visible — someone would have proposed weaning the country off housing subsidies several years before the worst excesses took place. Maybe not, but we'll never know because this ratio wasn't measured until it was too late.

A subsidy is a very heavy-handed, paternalistic, expensive, and economically distorting approach to public-sector management of a private marketplace. A measurement tool is a much lighter touch, which doesn't require anyone to do anything or cost

---

35. See Shiller, Robert, The Subprime Solution: How Today's Global Financial Crisis Happened, and What to Do About It (Princeton University Press, 2008).

anything. Therefore, moving forward, *Why The Heck* proposes that—unlike the past debacles of housing, energy and health care—the emphasis on managing this innovation economy should be on measuring it rather than attempting to subsidize it, and allowing the private and local government sectors to learn from those measurements.

There are a myriad of ways to measure innovation but, as of now, none directly account for the impact of innovation on the economy as a whole as measured by total domestic jobs created by innovation as a whole.

Specifically, I propose a new Innovation Index, which would tally the number/proportion of Americans employed by companies that did not exist one year, five years, and 10 years earlier, as well as their average wages as compared to others. This simple yardstick would allow governments, investors, and entrepreneurs to chart innovation, nationally and in specific locations.

At the same time that this yardstick would measure job formation in new companies, which correlates closely with innovation, it wouldn't disrupt the marketplace by pushing innovation to take place in specific sectors. In accordance with the philosophy of *Why The Heck*, measurement is the lightest possible government intervention, designed to encourage an activity by providing information about it, rather than distorting the marketplace.

There are several reasons this would be a valuable index for innovation measurement.

First, the interstate and intercity comparison itself would shine a bright light on the ability of states and cities to create a climate that encourages innovation. While not all new businesses would be technology businesses, and vice-versa, the composition of the new businesses is not that relevant. Both new business start-ups as a whole and technology start-ups in particular are good harbingers of a state's climate for innovation...if they grow and provide high-paying jobs. Just tallying the number of new businesses ignores how fast they grow. New business start-ups themselves are often driven by layoffs in existing businesses, and are predominantly sole proprietorships in personal service businesses that don't generate job growth, so start-ups by themselves can be a misleading statistic. Looking at employment figures in new business would be a much more important measurement.

Likewise, looking at wages paid to those employees would serve as an excellent proxy for the quality of the jobs. For instance, until recently one of the fastest-growing new industries was housecleaning, like Merry Maids. Clearly the creation of a new Merry Maid job does not have the same effect on the economy as the creation of a technology sector job.

Policymakers would have a field day determining not just what factors encourage business start-ups (they can do that with today's data), but rather what factors encourage the growth of new businesses offering high-paying jobs. It's the *growth* of those start-ups, as measured by the number and quality of jobs created by them, which is the critical factor in creating an innovation economy.

Second, where states and local governments are using taxpayer-financed dollars to attract start-up companies, describing the benefits in terms of job creation can help "sell" the concept of the subsidy to the voting, taxpaying public. Whether such a subsidy is or is not a good idea in specific situations is not my call. Having just railed against heavy-handed federal government interventions, I am wary of encouraging the same on the state or local level.

I'm going out on a limb here — and no one can refute me because there is no index to do so — to say that Massachusetts would finish first in this Innovation Index, probably by a considerable margin. I'd guess that today, about 15% of the Massachusetts private-sector workforce works for companies not in existence 10 years ago. Other states might study why, despite much higher-than-average levels of housing prices, per capita taxes, and whining, Massachusetts is so successful. Curiously, as noted in the comments, most of its neighboring states are notably unsuccessful. I could speculate on the reasons for the difference, but there will be plenty of opportunities for others to do that in a more data-driven manner once the actual data comes in.

Third, as a result of the bright light, and the focus on learning what works, the entire economy would become more innovation-oriented. No specific government intervention in the marketplace is needed. Vietnam, health care and *Casablanca* teach us that people manage and maximize what is measured. The most effective and least intrusive government policy in this area could be simply to measure — validly, understandably, and loudly — and then get out of the way.

COMMENTS

Comments on this section, though sparse because indexes are rarely controversial, were uniformly positive. No one can really oppose measurement of previously unmeasured variables. Interpretation of those measurements — and setting them up in a manner consistent with national priorities — is critical.

POSITIVE COMMENTS

Two people pointed out that there are many innovation indexes already, though none as elegant and understandable as this one. The closest is called the "New Economy Index." It does what this chapter describes, but only once every five years, and even then with some significant lag time and with too much complexity to be reduced to a single understandable variable. As a result, it attracts little attention, as is perfectly evident because you haven't heard of it and neither had I. Other similar "innovation indexes" are statewide, or track the stock price performance of certain companies deemed innovative.

One person commented that, while this index has a lot of simplicity and is easily understandable, the problem isn't a lack of innovation indexes. It's that there are too many of them. Many are statewide only, or published long after the fact. What he advocates is ONE number published fairly soon after each quarter or month, calculated the same way for every state, and sponsored by the Bureau of Labor Statistics (BLS) rather than some nonprofit somewhere.

Another comment noted that the BLS should stay out of it. They don't collect their data in a timely enough way. This data would have to be collected directly from payroll withholding.

Someone else wrote that a Boston newspaper columnist had opined on the importance of "rebranding" New England as an "innovation economy,"[36] because the region has been known instead mostly for its quaint village greens, maple syrup, lobsters, *Cheers*, and comical hats. (OK, I am exaggerating a little.) But the columnist had failed to note that you can't successfully "brand" a region as being innovative without any

---

36. Kirsner, Scott, "Let's Re-Define New England's Brand Image," *The Boston Globe* (December 27, 2008)

evidence of differentially more successful innovation other than anecdotes and a really good feeling about it.

I'd opine that while Massachusetts is extremely innovative, the rest of New England is not particularly so. Connecticut is so un-innovative that until recently the state ran its own venture capital company. Is there any business less likely to be managed well by the public sector than venture capital, other than maybe a perfume company or a Roller Derby team? Creating its own venture firm signals a level of desperation that would never be found in a state that does attract new businesses.

Right next door in Rhode Island, the major innovations took place centuries ago, with Slater's Mill, and also—I don't want to leave anything out here and risk understating that state's contribution to our economy—a red chicken. I'll stop there so as not to get three more states upset with me, but here's the bottom line: Outside of Massachusetts, New England is far from a hotbed of innovation.

That may just be my opinion, but that's exactly the point—it's my *opinion*. This Innovation Index could settle the argument. In the immortal words of the great philosopher Senator Daniel Patrick Moynihan, "Everyone is entitled to their own opinions, but not to their own facts."

PART SIX

Inquiring Minds Want to Know How to Win the $1 Million Prize

Even though you've finished reading *Why The Heck* and you have to admit at least some of the ideas are pretty stiff competition, you still think you have a chance to Win the Big Prize. But before you submit, you might want to really dig down, to know more about the odds of winning the prize and who you are competing against, to be assured that you are not wasting your valuable time. If so, you might want some in-depth statistics, so here goes:

1. The Marianas Trench in the Pacific Ocean is 33,302 feet in depth

Here are some other statistics, which may not be as in-depth but at least they are about the demographics of people like you and me who submit ideas on *Why The Heck*: 95% of us are between the ages of 32 and 58, making us 45 on average. Roughly 50% of us are male and 50% are female, making us hermaphrodites on average. 80% of us are college-educated, though I for one didn't get to class much.

Don't be discouraged from making a submission if you didn't go to college, though. 58% of us don't believe that graduating from college automatically makes you bright. For instance my freshman roommate — and we're talking Harvard here — thought that in the Southern hemisphere, they celebrate Christmas in July.

87% of us agree that's really dumb, but 76% of us also fail to notice that even if his "logic" were correct, they'd celebrate it in June. 0.01% of us[37] think it's hysterical to send him a Christmas card every summer.

While 2% of us develop our ideas by being extremely well-informed because we read *The Christian Science Monitor* online, the other 98% of us have never even heard of *The Christian Science Monitor*. By contrast, 96% of us read cereal boxes regularly, though only 0.01% of us ever successfully collected enough boxtops to send away for a neat prize.[38] 8% of us are Captains of Industry, but the other 92% have as much chance of making captain as Gomer Pyle, U.S.M.C.

97% of us can't remember our 9-digit zip code, having recognized that, Postal Service protestations aside, the mail nonetheless reaches us in time to get our 18 Chances to Maybe Have Already Won the Publishers Clearing House Gala Sweepstakes. Yet 74% of us can remember the name of Fred Ziffel's pet pig on *Green Acres*. When asked to interpret those seemingly incongruous statistics, 42% of us say "Go figure."

To be sure, the validity of much of this data is vulnerable to challenge, mostly because I made it all up except the part about the Marianas Trench, my college roommate and my neat prize

---

37. Me
38. Me again, and it busted.

busting, which I'm still pretty bent out of shape about. Fact is, I have no clue about the demographics of people submitting ideas on the site.

Nonetheless, I do know what the rules are because I made them up too. Unlike the statistics, though, you can hold me to these—but you should probably check www.whytheheck.com for updates. And listen carefully: Once the prize is won, it's over, unless someone (and it won't be me) sponsors another one. Here is the *Reader's Digest* Q&A version of the rules, followed by some fine print.

How can I win this money?

The $1 million goes to the single individual who comes up with the first idea originating on the www.whytheheck.com website BEFORE 3/31/11 that is subsequently adopted into federal policy. There is no subjectivity involved in choosing the winner. It is totally objective. To claim that money, you will have to track the "audit trail" between your idea and the policy. And you can't submit an idea you cribbed from somewhere or someone else like George Costanza. Google date-stamps everything. Your posting on www.whytheheck.com has to be the first mention anywhere of whatever idea you propose.

To win $100, all you need to do is submit an idea to www.whytheheck.com that gets into the next edition of *Why The Heck*. There is also no guarantee that there will be a next edition, though the fact that you are reading this edition doesn't hurt.

To win $20, all you need to do is to submit a comment to www.whytheheck.com that gets published in the next edition of *Why The Heck*. It can be supportive or challenging or intentionally or unintentionally funny.

Speaking of which, comments and ideas can be published anonymously if you prefer, but *Why The Heck* will never preface a comment with: "And you'll never believe what a dumb idea so-and-so came up with."

How do I know the money will be awarded?

Established law makes sure that I can't offer a prize and then not pay the prize money if the criteria for winning the prize are met. It's that simple. Not everyone can win the money, though. Members of my family are not eligible, and that's a good thing for you because my sister invented the ShopAmerica Gift Card (see Chapter Four), which could easily win. An old girlfriend also proposed an idea and then broke up with me because she noted that exception to the rule and thought that if she were on track to win, "You would marry me for your money."

How to generate "outside the box" ideas

Suppose you are reading a newspaper, in this case the *Washington Post* on May 21, 2009. You see a long article on the high cost of being poor, a topic also addressed twice by ideas in *Why The Heck*. You note that, in addition to the costs of poverty already addressed in *Why The Heck*, the poor pay dearly for low-quality fruits and vegetables because they do most of their shopping at corner stores, being unable to afford transportation to the generally more remotely located large supermarkets. The poor end up eating few fruits and vegetables, a diet that contributes to low health status in urban areas. Clearly, this is an economic and a health problem... but how to solve it?

Having read *Why The Heck*, you immediately think that cutting the middleman out of the process, could be the answer. You ask yourself, "How can high-quality fruits and vegetables be delivered directly into poor neighborhoods at low cost?" Farmers' markets generally offer very high quality produce at prices much lower than corner store prices. So farmers' markets could be one answer, but why wouldn't there already be farmers' markets in poor neighborhoods, if the demand is there?

Remember the two economists and the $10 bill on the sidewalk? It is not always the case that every business opportunity is already identified, every profitable niche filled. Still, let's assume that this opportunity has been considered by farmers and rejected as uneconomic. Let's assume instead that the market needs a little "noodge" to get started. How would you provide that noodge? Popular farmers' markets, limited by the number of spots allowed for vendors' booths, often have a long waiting list. Perhaps the answer is as simple as letting farmers who agree to participate in farmers markets in underserved neighborhoods go to the front of the waiting list for the more popular locations.

So there you have it...a newspaper article, a market inefficiency, some creative thinking, and *voila*—a brand new *Why The Heck* idea to improve the health and lower the cost of living for poor people...all at no cost to taxpayers. See how easy that was? I just did it. Now go ahead and try it yourself. It tastes like chicken.

PRIZE RULES

FIRST PRIZE

A single prize of $1 million ($50,000 per year for 20 years, starting six months after a winner is announced to provide time to check the audit trail) will be awarded for the first idea submitted before 3/31/11 originating on *Why The Heck* that is implemented on the national or state level, subject to the family member and contractual relationship exceptions below.

To qualify for the prize, the idea must *originate* on www. whytheheck.com. The prize will not be awarded for an idea that was supported or posted on www.whytheheck.com but originated elsewhere. This is a creative problem-solving site, not an advocacy site. There must be an "audit trail" linking the national or state policy change back to the *Why The Heck* posting.

In the event of a "tie" in that two ideas are included in the same piece of legislation passed by Congress, *Why The Heck* reserves the right to split the prize in any proportion, in its sole discretion, based on its interpretation of the importance of each idea. In no event is *Why The Heck* required to provide more than $1 million prize.

Owners of *Why The Heck* and family members, and those contractually affiliated with them or with *Why The Heck*, may win official recognition as the first to have an idea implemented in a federal law, but are not eligible for the cash prize. If indeed a family member — or someone contractually affiliated with a family member or with *Why The Heck* itself — is the first to have an idea passed into law or made into a regulation, the contest ends until and unless another sponsor endows another prize.

All reasonable ideas received that are deemed to fit the criteria of *Why The Heck* will be posted on the site, but *Why The Heck* reserves the right not to post those that *Why The Heck* considers in its sole judgment to be not in keeping with the tenets of *Why The Heck*, including but not limited to those requiring subsidies or mandates, and including but not limited to those that are unoriginal, impractical, unproductive, defamatory, plagiaristic, or lousy.

In case you missed it the first time, in no event is *Why The Heck* liable for more than the $1,000,000 total value of the first prize.

In the event of dispute, the decision of *Why The Heck* is final, subject to arbitration by the Boston branch of the American Arbitration Association (AAA), loser to pay arbitration fees and the winner's counsel fees. Anyone commencing such an arbitration against *Why The Heck* or any individual associated with *Why The Heck* will be required to post in advance an amount equal to the AAA's standard arbitration fees, refundable if the claimant prevails.

Microheckonomics: A new category with its own prize rules

A new category of postings has been established in which non-federal or non-state governmental initiatives may be proposed. Any such Microheckonomic initiative used in a subsequent book earns $100 as soon as the book sells 10,000 copies. Examples:

- The Yankees should sell Lou Gehrig jerseys to raise money for research on Amyotrophic Lateral Sclerosis (Lou Gehrig's Disease).
- The bases of coffee mugs and blenders should be designed to prevent water from accumulating in them in the dishwasher.
- Airplanes should board people without carry-on baggage first so they can sit down without clogging the aisles.
- Departments of Motor Vehicles should offer pagers just like restaurants do. Nearby businesses will pay for them, to get the extra foot traffic.
- Sports teams should scalp their own tickets on game day and donate some of the higher prices to charity. To fight childhood obesity and raise money, schools should keep their gyms open evenings and weekends for students, and finance the required extra staffing by selling off-peak memberships to the general public.

SECOND AND THIRD PRIZES

As mentioned, $100 will be paid for any reader-originated idea published in any subsequent edition of *Why The Heck*. In the event of similar ideas, the first date-stamped idea wins the prize or the prize may be split, in the sole judgment of www. whytheheck.com.

$20 will be paid for any comments published in subsequent editions.

Liability for disputes arising under this section is strictly limited to $100 and $20 respectively.

FOURTH PRIZE

An all-expense paid trip to the World's Fair in Syracuse, plus $1000 in Stimulus Stamps.

SUBSEQUENT PRIZE RULE REVISIONS

Prize Rules are subject to revision at any time. Not to worry, though. Anyone operating in good faith according to the rules in effect at the time they post is eligible for a prize under the rules in effect at that time. We can't change the rules retroactively, unlike when Superman made the earth spin backwards to keep Lois Lane from dying.

Why the Heck Aren't we Already Doing this Stuff

BIBLIOGRAPHY

If you think the ideas in this book are thoughtful but the writing isn't very funny, you should probably read something by Dave Barry. He is much funnier than I am.

If you think the writing in this book is funny but the ideas aren't very thoughtful, you should probably read something by Thomas Friedman. He is much more thoughtful than I am.

If you think the writing in this book isn't very funny and the ideas aren't very thoughtful, then you should probably not tell anybody.

Why the Heck Aren't we Already Doing this Stuff

ACKNOWLEDGMENTS

The first footnote in the introduction claims that the "large majority of sentences" in *Why The Heck* are "technically true," and this sentence is one of them. But that doesn't mean some other sentences aren't. For instance, the sentence listing four reasons I wrote this book. One of those reasons included "to meet women." That sentence was indeed technically true when I started this book but I subsequently found the love of my life, Mary Troxell, before *Why The Heck* went to press. She gets both the first acknowledgement and the biggest hug. Thank you for every minute, Mary.

And now, in the immortal words of the great philosopher Irving Berlin, on with the show. The Fearless Crew who edited in depth or otherwise contributed includes:

- Allan B. Ecker, who taught me that "which" and "that" are not the same word with different spellings, whose left brain kept my right brain firmly in check, and who has never been shy about telling me that I'm nowhere near as funny as I think I am;

- Mitch Greenberg, my next-door neighbor who fixed things in the house while I was fixing things in the book, and who has never been shy about pointing out my shortcomings in home maintenance;

- Larry Hilibrand, who pointed out a mathematical fallacy that everyone else had missed, like the guy who pointed out that a Matisse in MOMA was hanging upside down;

- John Hoffmire, who taught me about poverty economics and the cost of being poor;

- Mary Kukowski, who has never been shy about pointing out that I am nowhere near as original as I think I am;

- Debbie Levenson, who never withholds her excellent suggestions about this book even when I beat her at Boggle;

- Anne "Foodmill" Lewis, who convinced influential people to read *Why The Heck* before it was published, which turns out to be harder than writing it;

- Katherine "Sweets" Lewis, who invented the ShopAmerica Gift Card but unfortunately for her it is not a coincidence that her last name is "Lewis," and she is therefore ineligible for the cash prize, because this is the best idea in the book;

- Lindt, for providing both sustenance and comic material during my writing process, although I was later shocked, shocked to find that importation had been going on in here!

- Sam Lippe, who originated the Domestic Content Labeling idea in Chapter Five and says, "Don't forget to credit me with the proposal to subsidize the World's Fair in Syracuse. I'm very proud of that idea too";

- Josh Parker, who knows a ton about economics, editing and trivia, and most importantly, he and Marilyn Chambers, the star of *Behind the Green Door*, attended the same high school (note to Debby Parker: not at the same time);

- Joe Sacco, who taught me the ins and outs (mostly outs, of course) of dying in a hospital and who has never been shy about telling me that despite spending two decades in health care, I don't know anything about medicine;

- Bob Stone, who has never been shy about telling me anything. Don't tell him I said this but he is usually right;

- Kashmira Varga, who came up with a great idea even though she only came to this country as an adult not speaking a word of English. Yeah, I know. It's not enough that immigrants are stealing your jobs. Now they are stealing your $100 awards too.

In addition, the following people provided inspiration either directly or indirectly for ideas in *Why The Heck*: Edouard Aghion, Elisabeth Allison, Michael Barone, Michael Bell, George Bennett, Michael Blumenthal, Laurie Burgess, Lydia Chiang, David Coursey, Tom Day, Jerry Doyle, Cristy and Nathan Espiritu, Fathe and Ellen, Al Gore, Grammypool, Jerry Healy, Regina Herzlinger, Alex Howlett, Tom Keane, Michael Kolowich, Mary Kukowski, Jenkin Lee, Gordon Lish, Joe Malone, Greg Mankiw, Barry Manuel, Kai Maristed, Robby Moore, Howard Newburg, Barack Obama, Josh Parker, the Poolings and Morsels and Grublets (oh, my), Michael Quilty, Victoria Reade, Fred Reichheld, Darrell Rigby, Norval Rindfleisch, Mitt Romney, Sandra Schultz, Jordan Scott, Mike Shnayerson, Mary Sibal, Paula Silva, Squibbage, Joel Stern, Sweets and Fennis, Frederick Tremallo, Will Tobey, Robert Whaples, Kinsee Whiteaker, Montel Williams, Polly Whiteside, Juliana Zee and of course the Biscuit.

Why the Heck Aren't we Already Doing this Stuff

About the Author

Al Lewis is not related to the Al Lewis who played Grandpa Munster. He doesn't even look like Grandpa. Quite the contrary, he is so much taller than the other Al Lewis, or for that matter than most other garden variety human beings, that with surprisingly little makeup he could pass for Herman. Another important distinction is that this Al Lewis still has a pulse (52 bpm – not bad, hunh?). Finally, the other Al Lewis never had a blog as successful as www.whytheheck.com. *Why The Heck*, America's Marketplace of Ideas™, is the #1 economic idea generation blog in the country, having generated more national and local media coverage (under the original "Think Outside The Box" name, www.thinkoob.com) than all other similar blogs combined.

Why The Heck became the #1 economic idea generation blog partly because it has attracted almost a million visits and partly because there aren't any others. *Why The Heck* is unique not just in specializing in idea generation, but in offering a $1 million prize for the first idea to "graduate" from the website to national policy. Mr. Lewis also has a background in idea generation himself, having been widely credited with the invention of chronic disease management, which is now in use at almost every health plan in the country. His current "day job" is also in disease management, www.dismgmt.com. Previously, he taught economics at Harvard, and he has run several mid-sized health care companies, including one that didn't go bankrupt. He also believes in upholding the highest standards of integrity. So even if you google him until he's blue in the face, you won't find any priors. He also has a really cool ringtone.

Yet despite all these qualifications, he still can't get his kids to clean up their rooms.